GRUMMAN
ALBATROSS
A History of the Legendary Seaplane

Wayne Mutza

Schiffer Military/Aviation History
Atglen, PA

ACKNOWLEDGEMENTS

Throughout my research for this book, I was fortunate and honored to have met people who share my sentiments about Grumman's workhorse. Without their kind assistance, this book could not have "gotten up on the step," much less airborne.

Thank you Hal Andrews, Darrel Barfield, John R. Butler, Carl Damonte, Reid Dennis, Bill Devins, Darwin Edwards, Lawrence Feliu, Todd Falconer, Roger Ferguson, Jack M. Friell, Grant Hales, Tom Hansen, Dave Hansen, George Krietemeyer, John Lameck, Bob Lawson, Terry Love, Lennart Lundh, Lonny McClung, Dave Menard, Larry Milberry, Tom Neville, Robert North, Kirsten Oftedahl, W.M. O'Rourke, Dave Ostrowski, Jim Peel, Steve Penning, Chuck Pomazal, Matthew Rodina Jr., Marion Shrode Jr., Ralph Smith, Larry Trick, Mark Trupp, David Wendt, Kirk Williams, Nick Williams, and Ron Williamson. Credit is also due the Pterodactyl Society, the Air Rescue Association, and the Air Commando Association.

And last, but certainly not least, to my wonderful wife Debbie, who lovingly and loyally provided the encouragement and support to complete this project – and to our children: Jason, Joe, Joanna, Chris, Jeff, and Andy.

DEDICATION
This book is respectfully dedicated to the ALBATROSS crewmembers who lost their lives during the honorable performance of their duty.

Book Design by Robert Biondi.

Copyright © 1996 by Wayne Mutza.
Library of Congress Catalog Number: 95-71829

All rights reserved. No part of this work may be reproduced or used in any forms or by any means – graphic, electronic or mechanical, including photocopying or information storage and retrieval systems – without written permission from the copyright holder.

Printed in China.
ISBN: 0-88740-913-X

We are interested in hearing from authors with book ideas on related topics.

Published by Schiffer Publishing Ltd.
77 Lower Valley Road
Atglen, PA 19310
Please write for a free catalog.
This book may be purchased from the publisher.
Please include $2.95 postage.
Try your bookstore first.

CONTENTS

	Preface	3
Chapter 1	Pride of the Ironworks	4
Chapter 2	Specifications	8
Chapter 3	Improving the Flock	21
Chapter 4	Performance	23
Chapter 5	Sub Hunter	27
Chapter 6	The Air Force Albatross	30
Chapter 7	Coast Guard Goats	52
Chapter 8	Naval Aviation	68
Chapter 9	Worldwide Service	77
Chapter 10	"TT" Airlines	84
Chapter 11	Albatross Explorer	86
Chapter 12	Chalk's Legacy and the G-111	88
Chapter 13	Warbirds and Pleasure Craft	91
Chapter 14	Turbo Albatrosses	94
Chapter 15	Museums and Displays	96
	Line Schemes	102

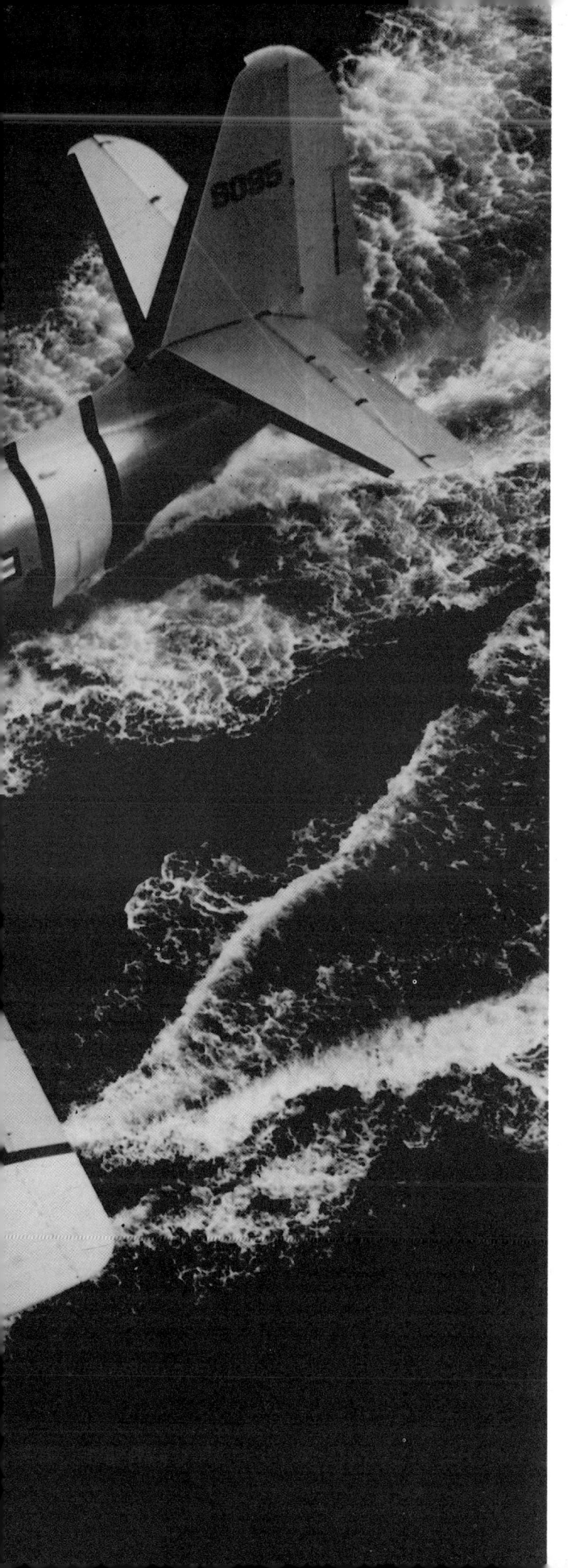

PREFACE

Though my background centers primarily around rotary-wing aviation, I have long harbored a fondness and admiration for the older radial-engine airplanes. Selecting one to research for a book was not difficult - not only is the Albatross a beautiful airplane with a unique and colorful past, but homage due the Albatross is long overdue. The original models were perfectly proportioned and all were incredibly versatile, reliable, and 'tough as nails.' This fascinating and worthy aircraft was aptly described in 1959 by George Eric Krietemeyer, Captain, USCG (Ret.):

"Most of us were introduced to the Albatross when we studied the English poet Coleridge's 'The Rhyme of the Ancient Mariner' in high school. We came to know the great white birds better during our summer cruises as they followed in the wake of the EAGLE, looking for leftovers from the culinary delights we were presented daily.

Those of us who have pinned on the 'Wings of Gold' as Coast Guard aviators have come to know and love a slightly different version of the bird. Its characteristics, both on and off the water, are very similar. Both live near the water and spend countless hours searching over the oceans for their livelihood. On the water they are both ungainly and not highly maneuverable. When taking off, they skitter across the surface, straining for the airspeed required to become airborne until, just at the right moment, they break the surface tension and laboriously claw the air for altitude. Once airborne however, they can fly effortlessly for hours on their missions, returning only for rest and nourishment when required."

In the hands of American military aircrews, and wearing international colors, the Albatross performed in stellar fashion. A veteran of two wars, the '16 became legendary, prompting tales that clouded the perception of fact and fiction. The 'old girl' was, after all, of Grumman lineage, earning it a place among the ranks of stalwart aircraft that brought their crews back despite overwhelming odds. Forgiving and tenacious as they were, Albatrosses didn't always conquer those odds - the price paid for the countless lives saved during three decades of yeoman service.

While I have endeavored to present a comprehensive treatise of the Grumman Albatross, these pages provide but a smattering of the many accounts that shaped the amphibian's abundant history.

Wayne Mutza

CHAPTER 1

PRIDE OF THE IRONWORKS

"Dumbo; Slobbering Albert; Goat; Clipper Duck" — oddly, all were endearments bestowed upon the Albatross by those who nurtured the amphibian, saw it through its military career, or shared in its continued success. Nearly half a century ago the Albatross embarked on a historical journey that earned it a place in the annals of aviation history.

Since 1929, Grumman's early years were spent designing and manufacturing floats for Navy aircraft and float planes. These beginnings evolved into a line of amphibians with the progenitor popularly known as the "Duck." First flown in 1933, the Duck was followed by the twin-engined "Goose" in 1937 and the smaller "Widgeon" in 1940, both of which served admirably during World War II. By war's end the Navy placed increased range and payload demands on its largest amphibian, the Consolidated PBY-5A "Catalina." As Grumman was producing its first post-war amphibian, the "Mallard," its engineering team set to work on a design to meet the Navy's requirements for a utility amphibian to replace the aging Catalinas. Incorporating the wealth of experience gained with early models and a continuous program of hydrodynamic research, the Grumman team conceived the Model G-64 as a continuation of the Goose design. The proposal for the larger new amphibian was presented to the Navy's Bureau of Aeronautics in mid 1944. The Navy accepted the proposal and awarded Grumman a contract for two prototypes designated XJR2F-1s. Grumman aimed for its first flight in 1946, however a number of delays resulted in the maiden flight accomplished on October 24, 1947. Water evaluations were underway the following month concurrent with the study of an anti-submarine warfare (ASW) version.

One month before the first Albatross took to the air, the U.S. Air Force was established and given responsibility for worldwide air rescue. This new mission had the Air Force shopping for a special amphibian and for Grumman, the timing could not have been better. The design aroused immediate Air Force interest and a mock-up was constructed in early 1948. The second prototype made its first flight in mid May as Air Force and Navy orders were drawn up.

By late summer, the Navy dropped its order for 32 PF-1A ASW variants in lieu of six production UF-1 utility transports. Conversely, the Air Force not only took over the canceled Navy contract but initiated orders for large numbers of SA-

The first XJR2F-1 prototype undergoes engine run-up tests at Grumman's facility at Bethpage, Long Island, New York. The aircraft is finished in dull aluminum paint and has a flight instrument probe attached to the wing tip. (Photo courtesy US Navy via Hal Andrews)

6 • GRUMMAN ALBATROSS: A HISTORY OF THE LEGENDARY SEAPLANE

Prototype number 82853 on its first flight on October 24, 1947. (Photo courtesy US Navy via Hal Andrews)

Nicknamed "Pelicans", one of the first two prototypes nears liftoff during initial water tests. (Photo courtesy US Navy via Hal Andrews)

Apparent in this view of number 82853 are the salient features which included a long sweeping afterbody and high-mounted engines. (Photo courtesy Grumman Corporation via Bob Lawson)

The first production Albatross, SA-16A number 48-588, is rolled out at Grumman's Bethpage plant after being fitted with triphibian gear for tests. The aft portion of the keel skid is in the lowered position. (Photo courtesy Grumman Corporation via Hal Andrews)

CHAPTER 1: PRIDE OF THE IRONWORKS • 7

Number 1240 was the first production UF-1G for the Coast Guard, seen here overflying Long Island, New York. (Photo courtesy USCG via George Krietemeyer)

For flight tests with the Naval Air Test Center, the second prototype was fitted with external fuel tanks. Number 82854 trundles up the amphibian ramp at NAS Patuxent River, Maryland. Note the crew member in the bow hatch. (Photo courtesy U.S. Navy via Hal Andrews)

Serial number 48-593 was the sixth SA-16A produced. Radio altimeter antennae are visible under the horizontal stabilizers. The football-shaped antenna atop the fuselage is a closed loop for an automatic direction finder. (Photo courtesy Grumman via Nick Williams)

8 • **GRUMMAN ALBATROSS: A HISTORY OF THE LEGENDARY SEAPLANE**

A Navy UF-1 shares ramp space at Grumman's Bethpage facility with Coast Guard and Air Force cousins: UF-1G number 2132 and SA-16A S/N 51-5289. (Photo courtesy Grumman via Jack M. Friell)

16As. The Coast Guard joined the program by placing an order for air-sea rescue amphibians under the designation UF-1G.

On December 30, 1949, the Naval Air Test Center (NATC) at Patuxent River received the first production Navy UF-1 BuNo. 124374 for BIS (Bureau of Inspection and Survey) trials. Aircraft BuNo. 124376 arrived for the tests on March 9, 1950 and both underwent intensive flight tests which lasted until February 13, 1951.

Seaplanes in the Navy date from January 26, 1911, and along with amphibians, were an important part of the Navy's aircraft fleet. There were drawbacks to the early types which hardly compared to sophisticated European aircraft.

The use of aircraft for air-sea rescue can be traced to the Battle of Britain in 1940 when numerous Luftwaffe and RAF flyers went down in the English Channel. The Germans actually pioneered the concept in 1939 when they modified fourteen Heinkel 59 float planes for air-sea rescue. During the war, Great Britain followed the example after losing nearly one fourth of their aircrew over the English Channel. The U.S. fortunately, had aircraft developed prior to the war that suited the air-sea rescue role. Though seaplanes were used during World War I, the first U.S. aircraft designed specifically for sea rescue was Consolidated Aircraft Corporation's PY-1 developed in the late 1920s for the U.S. Navy. Consolidated improved on their design with the PBY-1 "Catalina" developed in 1935. Although the Catalina required smooth seas for operations and had a limited range, it could slow-cruise to conduct searches ensuring its status as the mainstay of SAR during World War II.

For nearly two decades following World War II, seaplanes and amphibians served the U.S. Navy in utility and SAR roles. Air Force and Coast Guard Albatrosses ended their colorful careers during the mid 1970s finally falling victim to advanced technology which favors larger and rotary-wing aircraft.

CHAPTER 2
SPECIFICATIONS

Despite successive improvements and modification phases, all Albatross models retained their salient features which were slab sides, a deep "V" hull, high wings, unique undercarriage and compound fuselage and empennage curves. The metal skin of the airplane was flush-riveted except for the hull, wing floats, trim tabs, portions of the flaps and a narrow chordwise strip on each center wing section. Early production models mounted an AN/APS-31A search radar pod beneath the left wing. Since this limited radar searching on the right side, it was relocated from early on to the nose as a "thimble" fairing. Differing only in electronic gear installations, the Air Force SA-16A, Navy UF-1 and Coast Guard UF-1G were basically the same airplane.

The original Navy planning directive dated December 14, 1944, called for the installation of two Wright R-1820-74 engines. This was changed in the prototype to a pair of Wright Cyclone R-1820-76 powerplants rated at 1,425 hp each. All subsequent models were powered by R-1820-76A or -76B engines, with the exception of ten CSR-110s for Canada (Model G-231) and six UF-2s (Model G-262) for Japan. These derived their power from a pair of 1,525 hp Wright R-1820-82A engines which were distinguished externally by a cool-

This head-on view of HU-16B 51-7144 at the Robins AFB Museum shows the Albatross' deep hull angle. The dark patches on each side of the radome are flush interrogator antennae. (Photo courtesy Darwin Edwards)

Albatross prop hubs incorporated oil cooler "cuffs" fed from an independent oil supply tank inside the cowling. (Photo courtesy Darwin Edwards)

This view of 7144 reveals the upward tilt of the Wright engines. Unusual on this Vietnam veteran is the sea rescue platform attached below the rear cabin door. (Photo courtesy Darwin Edwards)

ing intake atop the cowling. Known as a "hot rod" engine of that period, the nine-cylinder engines featured single-stage two-speed superchargers and high-tension ignition systems. They were canted upward five degrees and mounted high for maximum clearance from water and spray. The eleven foot diameter propellers cleared the ground by eight feet with a four-foot clearance over the waterline.

Blended into the engine cowling over the wing were six exhaust stacks; three at the ten and three at the two o'clock positions. This meant that eighteen cylinders were exhausting through only twelve stacks. In early models, the clamps holding this arrangement together commonly broke apart, which in turn caused the unsupported stacks to break at the cylinder, creating a severe fire hazard. The condition turned flight mechanics and crew chiefs into scroungers for spare clamps until the problem was eventually corrected. The engines drove constant-speed three-bladed 43D50 Hamilton propellers which were controlled hydraulically. To position the blades through their controllable pitch and reverse movements, each prop had an independent oil supply in a semicircular tank located in the lower engine nose casing. This unique system had oil cooler "cuffs" mounted over the prop hubs. Propeller reversing was vital during water operations, however, detrimental to the early system was a moisture-sensitive prop-mounted electrical control switch. There were no safeguards against actuation and occasionally a prop slipped into reverse of its own accord. This caused a few anxious moments for the crew but was soon corrected by relocating the switch inside the prop hub. The relatively short props required for water operations produced an extremely loud roar at takeoff. An Albatross on JATO departure was said to be the loudest of any aircraft then in existence.

All Albatrosses were fitted with mounts for four JATO units, each of which produced 1,000 pounds of thrust for a 14 to 15

The first production SA-16A, serial number 48-588, had triphibian gear installed. The two-section skid can be seen fitted to the keel. (Photo courtesy Grumman via Dave Ostrowski)

CHAPTER 2: SPECIFICATIONS • 11

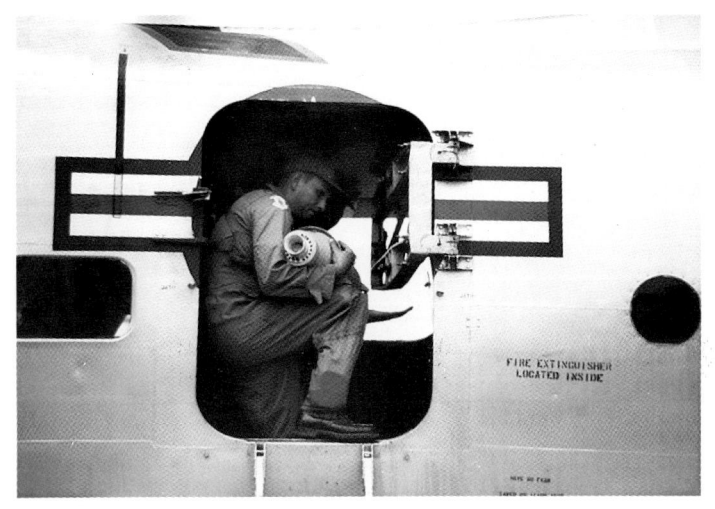

An Albatross crewman prepares to attach a JATO bottle to a Vietnam-based HU-16B. The inward-swinging racks allowed the bottles to be attached in flight. (Photo courtesy David Wendt)

During open sea training in 1967 at Drake's Bay near San Francisco, HU-16B number 51-7211 lost an engine which the crewmen tried to fix on the water. This view shows to good effect the pilot's escape hatch, prop hub, and engine work decks. (Photo courtesy David Wendt)

second duration. The attachments were incorporated into the port side cabin door and starboard side emergency hatch, both of which opened inward facilitating inflight JATO "bottle" mounting. JATO units were connected to electrical igniters and fired in salvos of two or four with a switch on the pilot's control wheel. The use of JATO reduced both land and water takeoff distances by forty percent.

The prototype had an empty weight of 18,200 pounds and a maximum weight of 26,000 pounds. The SA-16A, UF-1 and UF-1G had an empty weight of 20,800 pounds with a maximum weight of 33,000 pounds (29,500 pounds for water operations). The normal payload of production models assumed ten passengers, or twelve litter patients and one medical attendant, or 5,000 pounds of cargo, with a useful load up to 8,000 pounds in overload condition. The three types had a top speed of 240 mph and cruised at 150 mph. Their service ceiling was just under 25,000 feet. Technically, the Albatross featured a wingspan of 80 feet, a height of 24 feet, 5 inches and a fuselage length of 60 feet, 7 inches. A maximum range of 2,680 miles was possible with a full fuel load of 1,700 gallons. This was distributed among two internal main wing tanks having a total capacity of 675 gallons, two drop tanks of 100, 150, or 300 gallon capacity, and wing tip floats which doubled as 210 gallon capacity fuel tanks. All Albatrosses had these "wet" floats except the first six UF-1 and first 48 SA-16A aircraft produced. Underwing jettisonable fuel tanks were normally carried by all Albatross operators. The Air Force favored the 300 gallon MK 8 while the 150 gallon MK 12 and 100 gallon MK 4 were commonly seen on Navy and Coast Guard aircraft, all of which were attached to AN-MK 51 bomb racks.

HU-16B number 51-5292 wears an Air Force Outstanding Unit Award. Just forward of the wire antenna mast is a UHF TACAN antenna. (Photo courtesy David Wendt)

A crewman inadvertently hit the cockpit switch that jettisoned the external fuel tank. A nearby ignition source could have proven disastrous for the crew, which was in the aircraft briefing for a mission. (Photo courtesy David Wendt)

12 • GRUMMAN ALBATROSS: A HISTORY OF THE LEGENDARY SEAPLANE

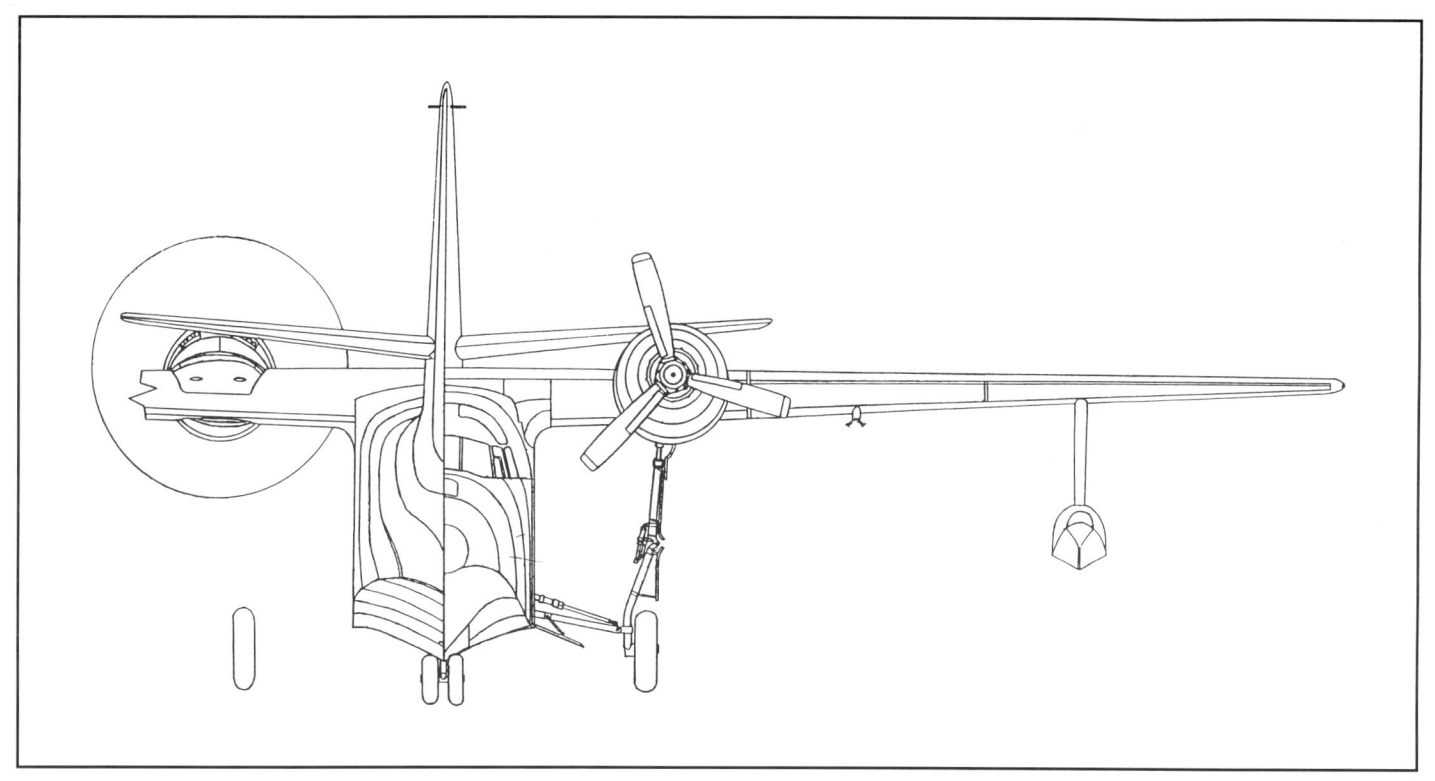

The Albatross was designed for prolonged operation from salt water bases with an airframe constructed of corrosion-resistant anodized aluminum alloy and zinc chromate primered throughout. The two-step aluminum alloy hull, which had a 7 foot, 11 inch beam, combined with an outriggered float on each wing tip to make the aircraft very stable in the water. Of the 466 Albatrosses produced by Grumman, the hulls for the 116th to 364th aircraft were built by the Plymouth Division of the Chrysler Corporation in Evansville, Indiana.

Albatross performance during land operations was maximized with the adoption of a wide track tricycle undercarriage. The width of the main gear, which was mounted beneath the engine nacelles, was 17 feet, 8 inches. Like its predecessors, the Albatross' articulating gear retracted hydraulically, flush into the fuselage. The Albatross was one of the largest aircraft without a steerable nose wheel. Instead, the unit, which had small weight and space saving high pressure tires, castored freely with directional control achieved

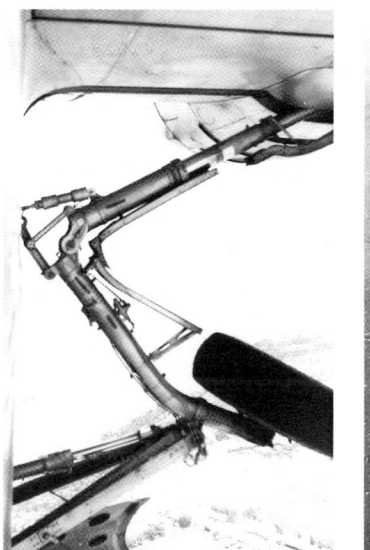

Left: Taken shortly after takeoff, this photo illustrates the intricate main landing gear folding mechanism halfway through its articulation cycle. (Photo courtesy Tom Hansen) Right: View looking forward of the main landing gear assembly. Note the smoke emitting from a hot brake following landing. (Photo courtesy Tom Hansen)

This view clearly outlines the gear wheel details as well as the hull step and step vents. The wheel covers seen here on the first Coast Guard UF-1G were quickly discarded to allow better fresh water washdowns after sea operations. (Photo courtesy U.S. Coast Guard)

CHAPTER 2: SPECIFICATIONS • 13

Coast Guard number 1293 was later converted to a UF-2G. The wing tip floats were one of the few areas of the Albatross that had external rivets. (Photo courtesy U.S. Coast Guard)

The Mark 8 fuel tank, seen here being mounted to a UF-1G, carried 300 gallons. (Photo courtesy U.S. Air Force via George Krietemeyer)

This view of Coast Guard number 1315 accentuates the Albatross' massive slab-sided fuselage. Note the bow pendant and "bubble" type cabin window. (Photo courtesy National Museum of Naval Aviation)

14 • GRUMMAN ALBATROSS: A HISTORY OF THE LEGENDARY SEAPLANE

Wooden JATO mockups affixed to the cabin door of UF-2G number 2127. Note the fold down step to the left of the "D" which provided access to the top fuselage walkway. (Photo courtesy the Grumman Corporation)

Below: "Front office" of the second prototype, number 82854. (Photo courtesy U.S. Navy via Hal Andrews)

by the use of rudder, brakes, reverse thrust and differential power.

To expand its versatility, an aberration unique to the Albatross was the triphibian gear installation. This system enabled the aircraft to operate on snow and ice while retaining its inherent competence on water and paved runways. Tests conducted with the system installed on the first production Albatross found no adverse affect on water operations. The Air Force ordered the system built into 145 of its SA-16As. Aircraft with serial numbers 51-032 through 51-7208 were factory-fitted with a reinforced mounting surface on the keel to accept a 15 foot two-section skid and shock strut. The forward 8 feet of the skid was affixed to the keel just aft of the nose wheel doors. Hinged immediately behind that was a 7 foot ski which ended at the main step, where it was attached to a combination shock strut and reaction cylinder. The aft ski could be extended downward 8 inches and locked. Both ski sections were 12 inches wide and the movable section was replaceable with a false keel to convert to amphibian operation. Completing the kit were shock-absorbing swiveling skids attached to the wing float tips. The skids were 33 inches long and 5 inches wide. Wing float tips were either fitted with mounts to accept triphibian skids, faired over, or incorporated

CHAPTER 2: SPECIFICATIONS • 15

The Albatross cockpit was a somewhat cramped but practical arrangement. UF-1, Bureau Number 137906, seen here in 1961, had a light gray instrument panel. (Photo courtesy the Grumman Corporation)

mooring rings (fore and aft). Later triphibian gear took the form of a down-sized version that comprised only the fixed ventral skid and removable float skids. The Air Force restricted Albatrosses operating in the triphibian mode to a maximum 29,000 pound takeoff weight. Canada's ten CSR-110s and two U.S. Navy UF-1Ls (LU-16Cs) were also completed as triphibians.

Externally, the Albatross lent itself to ease of maintenance. Its Wright engines and Hamilton props were common types with a good availability of spares. Integral with engine access doors was a nacelle work platform, with additional platforms for use on both sides of each nacelle, stored in the tail compartment. Ample walkways were provided on top of the aircraft.

The primary entrance to the aircraft was a three by four foot door at the left rear fuselage, commonly called a "dutch door." The top portion of this door swung inward and aft while the lower section dropped down and in. A boarding ladder, that attached to the base of the door, was stowed on the underside of the overhead cargo hatch. A sea rescue platform, better known as a rear boarding step, could be hung below the door to facilitate hauling people aboard while on the water. It was stored in the tail compartment and never used on Vietnam-based aircraft. If the Albatross faltered at anything, it was dropping rescue jumpers. The doorway was too small to allow heavily laden jumpers a good body position which is vital to parachuting safety. In addition, minimal use of the flaps for low speed stability created vortices that tumbled jumpers as they cleared the aircraft. A jump platform, static line cable, and bail out signal system were provided as standard equipment.

Opposite the cabin door, on the right side of the fuselage, was a two by three foot hatch. A larger hatch, which measured 5 feet, 3 inches by 4 feet, 10 inches, was located at the upper fuselage between the wing flaps. The overhead hatch was large enough to accommodate an Albatross en-

Navigator's position on the UF-2. (Photo courtesy the Grumman Corporation)

CHAPTER 2: SPECIFICATIONS • 17

Once the pilots were seated, they lowered this center hinged console. (Photo courtesy the Grumman Corporation)

Cabin view facing aft in the first Coast Guard UF-1G. On the aircraft's left is the "head" and washroom while the right compartment houses the APU. On the floor behind the seats are parachutes and a MK 7 life raft. (Photo courtesy the Grumman Corporation)

gine on a transport dolly or other bulky cargo. It was seldom used but checked carefully on preflight inspections since it was a structural part of the airplane — one the crew didn't want blowing off into the tail section. Incorporated into the hatch opening was a periscope sextant mount plus attaching points for the boarding ladder which provided access to the top of the aircraft. Above the pilot and copilot stations were escape hatches hinged at the outer edge. Five windows lined each side of the cabin, the middle of which were commonly replaced with an observation "bubble." A porthole was located aft of the cabin entrance door and another at the radio operator's position behind the copilot.

The horizontal stabilizer spanned 29 feet and had a 5 degree dihedral. The rudder and elevators were fabric-covered and de-icer boots along the wings and tail surfaces were standard equipment. A proportional rudder boost was provided for use during takeoffs, landings and single-engine operations. Split wing flaps were operated hydraulically while trim tabs, engine cowl flaps and oil cooler flaps were electrically operated. The outer leading edges of the wings were slotted the span of the ailerons for more positive low-speed control. Small leading edge spoilers were attached to each wing between fuselage and engine to ensure that stalls began at that point.

The aircraft bristled with various types of antennae indicating a full complement of communication, navigation, and radar equipment for day, night, and all-weather operation. As a result, it was frequently called upon to intercept and escort aircraft that experienced mechanical trouble.

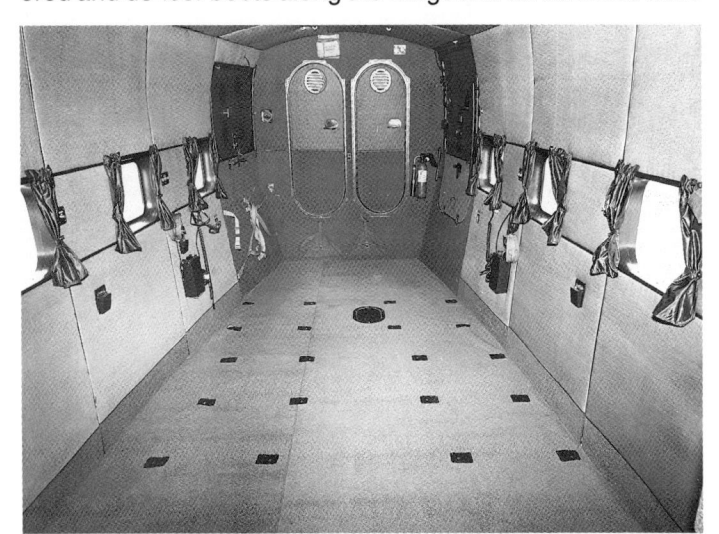

View facing aft in the cabin which is fully carpeted and soundproofed. Note the floor tie downs and details of the emergency escape hatch on the aircraft's right side and the "dutch" door on the left. (Photo courtesy the Grumman Corporation)

In this cabin view looking forward, the large cargo hatch in the top fuselage is visible. At the forward edge of the hatch is the life raft access. (Photo courtesy U.S. Navy via Hal Andrews)

18 • GRUMMAN ALBATROSS: A HISTORY OF THE LEGENDARY SEAPLANE

In this view looking forward from between the wheel wells, the interior of this UF-2 takes on a spartan appearance without carpeting and soundproofing. (Photo courtesy the Grumman Corporation)

Attached to the nose and encircling the radome was a bow pendant for attaching mooring lines, tow lines or a sea anchor. Just aft of the pendant was a bow hatch over a bow compartment where an anchor, mooring hooks, windshield fluids, and oxygen bottles were stored. Bilges in the compartment could be filled to hold the aircraft's nose down in the water.

Moving aft from the bow compartment through a crawlspace led to the flight deck with crew seats of the more comfortable long-range type. A crowded but practical cockpit arrangement made the Albatross a "pilot's machine." Typical of amphibian and flying boat design, the vital controls were arranged overhead to provide the shortest distance to the high-mounted engines. Once in their seats, the pilots lowered a hinged center instrument panel between them. The radio operator was positioned behind the copilot on the right side.

Moving aft through a bulkhead came the forward cabin where the navigator's position, also on the right, comprised a desk, swivel seat, and nav/search radar. The aft cabin section featured 145 square feet of floor area, 568 cubic feet of volume, and incorporated 48 tie down points in the floor and sidewalls. A wide center of gravity range allowed loading flexibility. For medical evacuation, the cabin could accommodate 12 litter patients or 11 Navy-type stretchers. A fold-down medical attendant's seat was hinged to the right wheel well. Albatrosses not built as triphibians (S/N 51-7209 and subsequent), could accommodate only eight litters but had a refilling station for oxygen bottles used for casualties. The cabin was also equipped for numerous combinations of ten passenger sets. Also stored in the cabin were parachutes, JATO bottles, food storage lockers and three life rafts; two were 7-man MK 7s stored in the aft cabin and a smaller automatic 4-man MK 4 was stored just forward of the overhead cargo hatch. It was releasable from inside the cabin or outside the aircraft through its own hatch. Special air rescue kits could also be carried and air dropped to provide flotation, shelter, and medical supplies for up to forty sea survivors.

A bulkhead separated the aft cabin from two side by side compartments; the right side housed the auxiliary power unit (APU) which used fuel from the right main tank and delivered up to 10 hp. It drove a generator to supply power on the ground when external power was not available and also in flight if the main-engine generator failed. The compartment also con-

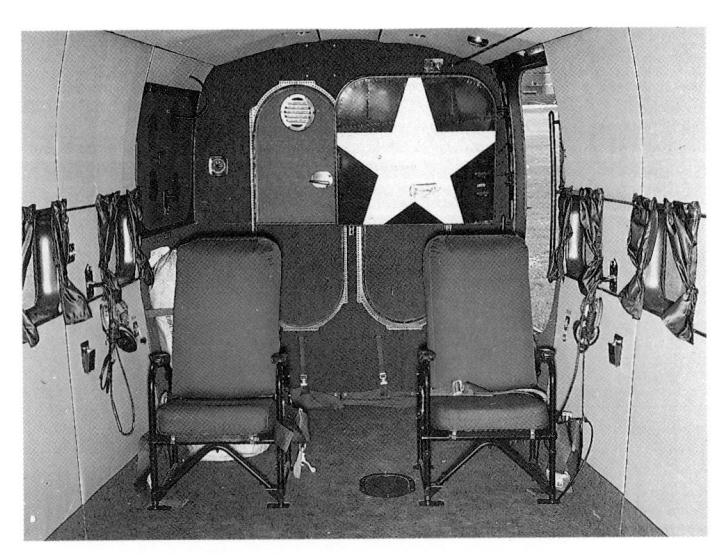

This Navy UF-1 has two observer's seats in the rear cabin. Note how the top half of the two-section cabin door swings inward. The large number of retaining pins securing the immense overhead cargo hatch are visible. (Photo courtesy the Grumman Corporation)

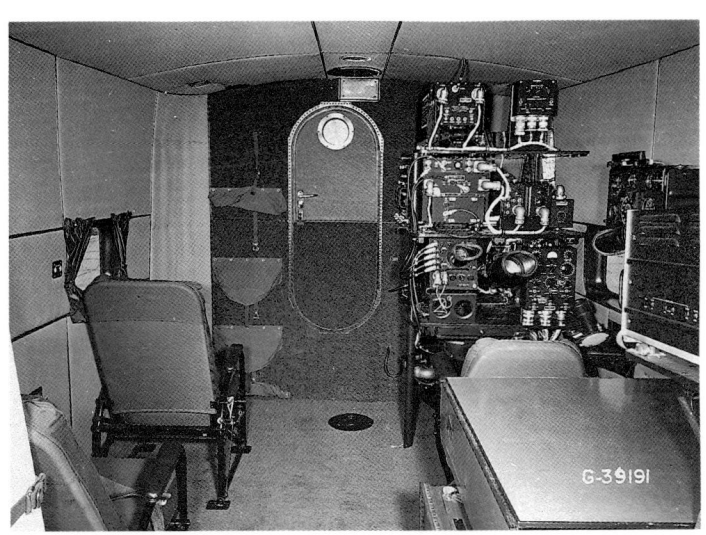

UF-1 cabin interior showing bulkhead door to flight deck and navigator's station. (Photo courtesy the Grumman Corporation)

tained two 17 GPM bilge pumps with four hoses. The left side compartment was furnished as a lavatory with a chemical toilet and fresh water basin.

Furthest aft was the tail compartment which contained the hydraulic accumulator and storage for the sea boarding platform. Two parachute flares stowed there in fixed containers were fired electrically from the cockpit. Additional Albatross equipment included a complete oxygen system for the crew and passengers as well as a portable electric fuel pump and a "Very" flare pistol.

U.S. AIR FORCE/U.S. NAVY ALBATROSS COLORS

NAME	FEDERAL STANDARD No.
Dark Sea Blue	15042
Insignia Blue	15044
Black	17038
White	17875
Yellow-orange	13538
Insignia red	11136
Fluorescent Red-orange	633
International orange	12197
Camouflage Gray	36622
Aircraft Gray	16473
Light Gull Gray	36440
Seaplane Gray	16081
Engine Gray	36081
Medium Gray	15237

Note: Numbers beginning with "1" are flat while those beginning with "3" are gloss.

GRUMMAN ALBATROSS: A HISTORY OF THE LEGENDARY SEAPLANE

CHAPTER 3

IMPROVING THE FLOCK

In 1955, Grumman developed an Albatross conversion plan, designated G-111, which satisfied specific operational requirements of the Air Force Air Rescue Service. Up until that time, one glaring drawback of the Albatross was its adverse performance during single-engine operation, which was compounded by gradual weight increases. Beginning with aircraft S/N 51-7200, first flown on January 11, 1956, a total of 86 SA-16As initially underwent the Air Force-sponsored modification program. Conversions were accomplished during IRAN (Inspect and Repair as Necessary) programs at Grumman. The first production SA-16B was flown on January 25, 1957, and a total of 241 Air Force SA-16As were eventually upgraded to SA-16B standard.

The Navy followed suit by initiating conversion of its UF-1 fleet to UF-2 standard in 1957. U.S. Coast Guard UF-1Gs brought up to USAF SA-16B and USN UF-2 standard were identified as UF-2Gs. This program encompassed Grumman Design Numbers G-234, G-270, and G-288, which were differentiated only by minor systems installations. During the period November 5, 1958, through January 15, 1959, the Naval Air Test Center conducted a preliminary evaluation of a Coast Guard UF-2G. The first Navy UF-2 took to the air on January 13, 1959, with the type placed in service on January 30th. The only major difference between the Navy and Coast Guard variants was a completely revised electronics installation in the UF-2G.

Improved performance was achieved through aerodynamic refinements which involved a number of structural changes. The wing area was increased from 883 square feet to 1,035 square feet with the addition of 70 inch panels added immediately outboard of the engine nacelles. 30 inch extensions added to each wing tip stretched the wing span to 96 feet, 8 inches, an increase of nearly 17 feet over earlier models. To compensate for the longer wings, the aileron span was increased by 60 inches and a geared balance tab added. The new wing length also had the wing floats moved outboard 70 inches for greater stability on the water. The flaps were redesigned with the insertion of a 70 inch adapter section and revision of the outboard portion which decreased the total area. Cambered leading edges replaced wing slots ensuring lower stall speeds and chordwise high pressure de-icing boots replaced the previous arrangement of spanwise boots.

Horizontal tail surfaces were enlarged by adding 12 inch squared off tips resulting in a span of 31 feet. The chord was extended forward by a new leading edge. An 18 inch fiber-

Framed by an Albatross wing and float support in the foreground, three Navy UF-1s undergo overhaul and conversion to UF-2 standards at the Grumman plant. Bureau Numbers of the trio are 137907, 142358, and 141283. (Photo courtesy the Grumman Corporation)

glass cap was added to the tail fin, enclosing the once visible antennae, and extended the airplane's overall height to 25 feet, 10 inches. Where the fuselage and rudder support structure were moved rearward eight inches, the rudder and its hinge line followed.

An internal centrally located gust lock system was added and aerodynamic drag was reduced by flush mounting high-drag antennae. The AN/ARA-8 VHF homing antenna took the form of a trapezoidal panel on the wing float supports.

Whereas original production machines experienced descent with loss of an engine, the modified version could not only maintain single-engine flight, but climb as well. Vast improvements of the improved Albatross included an increase in cruise speed, lower stall speed, and range extended beyond 3,000 miles. The airplane's maximum gross weight increased by 5,000 pounds.

It was discovered that the same program intended to extend the life of the Albatross may have actually shortened it. Tests by the Navy's Research Center revealed that a series of holes, drilled in the wing spar during modification, could cause the wings to fail at 19,000 hours. As a result, a flight restriction of 11,000 hours was placed on the airframe, forcing the Albatross into what many felt was premature and unnecessary retirement. There is speculation that the cap on flight hours was conveniently imposed at a time when the military actively sought a blanket transition to jet aircraft, thereby eliminating the need for radial engine training and maintenance. Perhaps the uncooperative Albatross lived beyond its years.

As part of the revised designation system instituted in 1962, the SA-16B became the HU-16B, the Navy UF-2 changed to HU-16D and HU-16E identified Coast Guard UF-2Gs.

Owing to the success of the modification program was an incident involving a HU-16B of the USAF 33rd ARRS based at Naha Air Base, Okinawa, during the mid 1960s. After the loss of one engine over the ocean, midway between Hawaii and Okinawa, the Albatross flew on to its destination — a distance of 550 miles which required a five hour flight.

Serial number 51-7200 was the first Albatross modified to "B" Model configuration. This perspective emphasizes the increased wing span, increased chord horizontal stabilizers, heightened fin and improved de-icers. The SA-16B is seen here two weeks after its initial flight. (Photo courtesy U.S. Air Force Air Combat Command)

This front view of the prototype SA-16B shows the stretch of the extended wings. (Photo courtesy U.S. Air Force via Hal Andrews)

CHAPTER 4
PERFORMANCE

Though the original Navy requirements specified a utility aircraft, the Albatross saw extensive service in search and rescue and performed equally as well for cargo and personnel transport, training, medical evacuation, patrol, escort, special operations, command and control, and anti-submarine warfare. To fulfill these mission requirements, the Albatross lent itself to a multitude of configurations.

As the largest and most powerful of the Grumman amphibian family, the Albatross was of similar configuration to the JRF "Goose." Ungainly in appearance when perched on its unusual landing gear, the slab-sided craft hardly evoked visions of its graceful avian namesake. Nevertheless the Albatross proved to be a superb combination of performance, ruggedness, and versatility unmatched by any other amphibian. Its structural integrity and excellent hull design formed the cornerstone that ensured its reliability for missions under conditions that would have guaranteed the demise of weaker aircraft. The Albatross notwithstanding, had its limits and often paid the ultimate price. Durability became a byword for the Albatross — they have been landed wheels-up on concrete and flown away after temporary patching. In 1956, one example was skillfully landed among 15 foot waves to rescue a B-26 bomber crewman. The unforgiving seas denied any attempt at takeoff so, the pilot turned sea captain and taxied his aircraft 98 miles back to base at Okinawa — an unofficial record. Another Okinawa-based Albatross performed an outstanding feat by flying 550 miles on one engine during a ferry flight. When the failure occurred over the Pacific Ocean, the airspeed dropped to 110 knots and the amphibian was nursed to Naha Air Base.

Equally impressive was the Albatross' ability to withstand weather extremes. During daily use in arctic and antarctic regions, where temperatures never exceeded -15 degrees fahrenheit, engines were started without problems. Conversely, a pair of SA-16As served an Air Rescue flight at Dhahran, Saudi Arabia, where temperatures above 100 degrees and sandstorms were daily occurrences.

Improvements in the Albatross allowed an endurance of nearly 20 hours and a maximum range of more than 3,000 miles which made it ideally suited for patrol and search missions. It possessed excellent short field takeoff ability from fields and bodies of water normally inaccessible to aircraft of its size. This capability was further enhanced by the use of JATO units. Early models could clear a 50 foot obstacle at sea level at 28,000 pounds gross weight in 2,500 feet, which was significantly reduced to 1,500 feet with JATO. The im-

SA-16A number 52-136 was one of five Albatrosses transferred to Spain under MDAP in 1954. Here, it performs a JATO takeoff while still in USAF markings prior to delivery. Each JATO bottle provided 1,000 pounds of thrust. (Photo courtesy U.S. Air Force via Dave Ostrowski)

Prior to conversion to a "B" configuration, SA-16A number 51-5282 was triphibian-equipped. Here, in early USAF Rescue markings, it rests on its keel preparing for takeoff from an ice field. Number 5282 was the last operational Air Force Albatross, which made a world record altitude flight before being placed on display at the USAF Museum. (Photo courtesy J.B. Chessington via Dave Menard)

CHAPTER 4: PERFORMANCE • 23

Photographed from another Albatross, a HU-16B makes an open sea takeoff run in 1970. (Photo courtesy David Wendt)

proved version could be airborne over the same obstacle at 36,000 pounds gross weight in 4,250 feet which JATO reduced to 2,550 feet.

It was in the flying boat role that the Albatross truly came into its own. It compensated for whatever awkwardness it displayed on land by transforming into a thing of beauty on the water. And it was then that Albatross pilots distinguished themselves as a flying fraternity of vessel operators, whose prudent members learned to "read" the sea and quickly develop a "feel" for the skillful balance of power and control they coaxed from their machine turned seaplane. Equally important was their ability to decide when it was possible to land or take off safely. Albatross pilots agree that open sea operations were a hazardous undertaking in a decidedly hostile environment. Bearing testimony to that sentiment are a number of Albatrosses that lie on the ocean's bottom. Among the challenges Albatross pilots met on the open sea were nonexistent visual references, 40 to 50 foot sea swells that created mammoth troughs and crests, and "confused seas" that interspersed calm spaces and criss-cross wave patterns.

The Albatross was a much gentler and more forgiving creature on the water than earlier flying boats and amphibians. Its superbly designed hull dampened turning tendencies on takeoff and, drawing four feet of water, reduced dangerous "porpoising." Critical during water takeoff was "getting up on the step," that phase when the aircraft transitioned from boat to hydroplane. The sharp angle of the hull and powerful engines enabled the Albatross to withstand rough seas and make quick takeoffs, even in smooth water. On the other hand, a lesser angle allowed quicker liftoff from calm water with heavy loads. Consequently, the Albatross hull was a compromise, as everything else in aircraft design.

Although it was designed to operate from four foot seas, the Albatross has successfully taken off and landed in eight foot swells with and without JATO. Pilots preferred slightly choppy seas for takeoff since a takeoff run on smooth water created an adverse suction effect. Not only did JATO help to overcome this effect but it spared the airplane and crew from a pounding takeoff run. During takeoff attempts, there was an ever present danger of caving in the radome or, especially, the nose gear doors.

During the displacement phase, when the aircraft is afloat, a number of techniques were available to the pilots for control. Reversible props allowed the Albatross to back up at a moderate speed and the engines could be used for maneuvering. Water rescues were usually accomplished using the engine opposite the survivor in forward, idle, and reverse. For positioning, the onboard sea anchor could be used or the landing gear lowered for greater drag. In addition, the Albatross could water taxi with one engine by means of a drogue on the power side, or be towed by a ship from the bow or stern.

With its landing gear down for added stability in the water, SA-16A number 51-7195 takes survivors aboard. Victims taken from ships or in accessible land areas were usually floated to the aircraft by raft. The crewmen are using the sea rescue platform. (Photo courtesy U.S. Air Force via Dave Ostrowski)

The Albatross distinguished itself in the record-setting arena as the holder of nine world class amphibian records achieved by crews from the U.S. Coast Guard, Navy and Air Force.

On August 13, 1962, Coast Guard CDR W.C. Dahlgreen and CDR W.G. Fenlon established two speed records by flying a 621.5 mile course at an average speed of 232 mph with payloads of 2,205 and 4,409 pounds.

On September 11, it was the Navy's turn when LCDR D.E. Moore took the UF-2G to 29,460 feet, setting a new world record for amphibians carrying a 2,205 pound load. The same day LCDR F.A.W. Franke flew the Albatross with a 4,409 pound load to 27,380 feet. Three days later, LCDR R.A. Hoffman set a new world speed record over a 3,107.5 mile course by flying at an average speed of 151.4 mph with a 2,205 pound payload.

Coast Guard CDR W.G. Fenlon set a new straight line distance record without payload with a flight from NAS Kodiak, Alaska, to NAS Pensacola, Florida, on October 24, 1962, a distance of 3,571.7 miles.

Coast Guard Albatross number 7255 was used for these record flights which originated on land and ended on water for FAI (Federation Aeronautique International) amphibian certification. Except for the distance record, all flights were flown from Floyd Bennett Field NAS, New York.

The Air Force got into the act in 1963 with one of their HU-16Bs, S/N 51-7211 of the 48th ARRS. On March 19th, Capt. G.A. Higginson piloted the Albatross over a 621.4 mile course carrying a 11,023 pound payload flying at an average speed of 153.7 mph. Carrying the same payload, Capt. H.E. Irwin climbed to a new record height of 19,747 feet the following day.

An SA-16A leans on its starboard wing tip float as it negotiates a tight water turn. Albatross pilots quickly developed a healthy respect for the sea and learned to "read" its many features. (Photo courtesy U.S. Air Force Air Combat Command)

CHAPTER 4: PERFORMANCE • 25

The first production Albatross undergoing open sea tests to prove that its triphibian gear does not affect its water handling characteristics. The upper wing section between the engine nacelles is painted yellow-orange. Note the wing slots and absence of an anti-glare nose panel. (Photo courtesy Smithsonian Institution)

On July 4, 1973, the last Air Force Albatross, S/N 51-5282 of the 301st ARRS, rolled down the runway at Homestead AFB, Florida, with Lt. Col. Charles Manning at the controls. Nicknamed "Chuck's Challenge" after the Colonel, the HU-16B was on its final mission, that of surpassing the existing altitude record for amphibians, set by an Albatross eleven years earlier. It accomplished the mission when it climbed to 30,700 feet, a record, like the others, that stands today. The Albatross landed off Watson Island in Biscayne Bay where it was taxied ashore and greeted by Grumman officials, among them Grover Loening, the original builder. A few days later, Number 5282, a veteran of the Korean and Vietnam wars, was flown to the Air Force Museum at Wright Patterson AFB, Ohio, where it rests in a place of honor.

CHAPTER 5

SUB HUNTER

The combined Grumman/Navy study of an anti-submarine Albatross initiated in early 1947, reached fruition some fourteen years later. Based on the improved SA-16B, the ASW Albatross retained all of its well-proven capabilities while offering this special function which enabled it to detect, identify, track and destroy submarines. Construction of the type was under way in 1960 with the first example flown on May 11, 1961. By March 1964 a total of 36 units were produced for delivery to foreign governments. Norway was the primary customer with orders for 18 aircraft followed by Spain (7), Chile (6), Peru (3), and Indonesia (2). Eventually, the government of Greece would procure 13 second-hand ASW Albatrosses, 12 from Norway and one from Spain. Initially labeled SA-16B/ASW, they were redesignated SHU-16B in 1962. Prior to delivery to Norway, the third and fourth SHU-16Bs produced were assigned to the U.S. Navy's VP-31 for evaluation trials.

Two major structural modifications were made to adopt the HU-16B for the ASW role. The keel was extended forward of the nose gear doors and the nose modified to accept a bulbous radome which contained AN/APS-88 search radar and a glide slope antenna. In addition, the lower aft tail section was altered for the installation of a retractable magnetic anomaly detection (MAD) boom. Both modifications extended the aircraft's overall length by only one foot.

Prominent among the myriad of electronics added to the sub hunting Albatross was wide sweep angle and long-range radar. This variant was further identified by numerous additional antennae located mainly on the fuselage spine and wing tips. Carried inside the aircraft were an underwater sound signal dispenser with 30 MK 50 or MK 57 signals and 20 MK 25 marine markers and launcher. A sonobuoy launcher pod and 16 sonobuoys were stored in the aft fuselage. The electrically fired pod could be attached in flight utilizing the JATO mounts on the main cabin door.

This early photo was retouched by a Grumman artist to illustrate the proposed ASW design during the 1950s. The MAD boom is shown extended. (Photo courtesy Collections of the Aviation History Branch, Naval Historical Center)

SHU-16B number 51-070 is fully loaded in the ASW mode. The sonobuoy launcher is attached to the JATO mounts over the cabin door and a torpedo and depth bomb are carried under the port wing. ASW Albatrosses were delivered under MAP/MDAP in USAF markings and serial numbers. This aircraft was one of 18 delivered to Norway. (Photo courtesy Collections of the Aviation History Branch, Naval Historical Center)

External armament stores were hung from 600 pound capacity Aero 15C combination bomb rack and rocket launchers, or two 15C racks and two 2,000 pound capacity 65A bomb racks. Various combinations of munitions could be carried on these racks depending on tactical requirements. Besides jettisonable fuel tanks, external stores included MK 43 torpedoes, MK 54 or MK 101 depth bombs, and 5 inch LAU-10/A "Zuni" high velocity aerial rockets (HVAR). A third Aero 65A bomb rack under the starboard wing carried a AN/AVQ-2C searchlight.

The SHU-16B's empty weight was 24,200 pounds, and fully loaded, weighed in at a hefty 35,500 pounds. This weight, coupled with the usual absence of external fuel tanks, held the aircraft's range to 1,125 nautical miles. Besides a crew of six, this variant accommodated additional crews for inflight ASW training.

One of Norway's ASW Albatrosses undergoes water tests prior to delivery. (Photo courtesy the Grumman Corporation)

ASW SA-16B/SHU-16B PRODUCTION

SERIAL NUMBER	RECEIVING COUNTRY	SERIAL NUMBER	RECEIVING COUNTRY
49-097	Chile	51-5288	Norway
49-099	Chile	51-5300	Norway
49-100	Chile	51-7147	Spain
51-014	Chile	51-7148	Spain
51-024	Chile	51-7165	Spain
51-038	Peru	51-7167	Spain
51-040	Norway	51-7170	Spain
51-041	Peru	51-7172	Spain
51-044	Norway	51-7174	Peru
51-048	Norway	51-7177	Norway
51-050	Norway	51-7183	Norway
51-060	Norway	51-7190	Norway
51-068	Norway	51-7191	Chile
51-069	Spain	51-7196	Greece
51-070	Norway	51-7202	Norway
51-474	Norway	51-7203	Norway
51-5281	Norway	51-7204	Norway
51-5283	Norway	51-7207	Norway

CHAPTER 6

AIR FORCE ALBATROSS

With the lessons learned from the massive and intense air campaigns of World War II fresh in their minds, U.S. military leaders addressed the matter of air and sea rescue during the post war period in typical fashion; the Army Air Forces believed its rescue capability should be expanded while the Coast Guard, supported by the Navy, contended that air and sea rescue was traditionally their responsibility. A compromise, accomplished through liaison with the Coast Guard, gave the AAF Air Transport Command the responsibility for search and rescue over land and along ATC routes. This proposal led to the establishment of the Air Rescue Service in March 1946 followed by assignment to the ATC on April 1st.

After the U.S. Air Force was established on September 18, 1947, it became responsible for worldwide air rescue operations. Air Force planners were then faced with an urgent requirement to equip rescue squadrons with amphibious aircraft. Though originally intended for the U.S. Navy, the Albatross design quickly gained favor with the Air Force, who became the primary buyer of the aircraft. The first contract was approved on May 12, 1948, and the first production Albatross (48-588) made its maiden flight on July 20, 1949.

An initial production batch of 32 PF-1As for the Navy was canceled and the contract amended allowing the Air Force to purchase them as SA-16As. An initial 32 units were delivered that year and another 11 in 1950. Albatross production for the Air Force peaked during 1951 with 225 aircraft. Two were built during 1952 for a USAF order and passed on to foreign governments as part of the MAP program.

When deliveries ended in December 1953, the Air Force had accepted a total of 293 SA-16As. Ordered as SA-16As, 15 aircraft bearing USAF serial numbers 52-121 through 52-135 were transferred to the Coast Guard during production and delivered as UF-1Gs numbered 2121 through 2135. In addition, 37 SA-16As in Air Force service were later transferred to the Coast Guard. Another ten machines, assigned Air Force serial numbers 60-9301 through 60-9310, were delivered to the Royal Canadian Air Force as CSR-110s. A number of SA-16As were configured as triphibians for arctic duty. A total of 241 examples would eventually be converted to SA-16Bs which were redesignated HU-16Bs in 1962.

Following its introduction to the Air Force in 1949, SA-16As were assigned to the Air Rescue Service of the Military Air Transport Service (MATS) where they replaced aging OA-10A "Catalinas." They also supplemented, and eventually replaced, fixed-wing rescue aircraft such as Boeing SB-17s and SB-29s as well as Douglas SC-47s. History has shown that military personnel never seem content with given aircraft

HU-16B 51-043 of the 56th ARS at Otis AFB, Maryland rests on frozen Mud Lake, Goose Bay, Labrador in February 1959 where it practiced ski landings. (Photo courtesy Carl Damonte)

30 • GRUMMAN ALBATROSS: A HISTORY OF THE LEGENDARY SEAPLANE

Stripped of its engines and exposed to the elements, number 51-7195 rests at the MASDC facility in April 1976. (Photo courtesy Nick Williams)

With JATO firing, HU-16B number 51-7178 breaks free of the water. (Photo courtesy Tom Hansen)

names, no matter how fitting, and the Albatross was no exception. Before long the Air Rescue community had nicknamed the amphibian "Dumbo" after Walt Disney's graceful flying elephant. Since its inception, the Air Rescue Service was organized into numbered squadrons, each with a headquarters and four flights lettered A through D — Flight A was normally based with headquarters while the rest were detached. The first SA-16As to serve abroad were serial numbers 49-074 and 49-075 sent to Flight D, 7th ARS at Dhahran, Saudi Arabia, the same month the Korean war erupted.

At the outbreak of the Korean war on June 28, 1950, two SAR units served the Far East Air Forces (FEAF); the 2nd and 3rd Air Rescue Squadrons (ARS). The 3rd, which bore the brunt of air rescue operations in Korea, was headquartered at Johnson Air Base in Japan (Flight A) with detachments at Misawa (Flight C), Yokota (Flight B), and Ashiya (Flight D). One month after the war began, three SA-16As were assigned to the 3rd ARS to enhance the unit's rescue capability and test the Albatross' mettle. The anticipation was short lived with the rescue of a Navy Ensign by an Albatross one week later. By Fall, SA-16As were operating from bases in Korea including K-2 Taegu, K-3 Pohang, K-16 Seoul, and K-24 Pyongyang. In November 1952 the 3rd ARS was upgraded to group level and augmented in December by the 2nd Air Rescue Group.

Throughout the war SA-16As flew constant daylight patrols over Korean waters, including the Tsushima Straits. The SA-16A's 12 to 14 hour extended endurance allowed it to

HU-16B 51-7194 is streaked with oil after it lost its number one engine over Vietnam on November 14, 1966. (Photo courtesy Tom Hansen)

CHAPTER 6: AIR FORCE ALBATROSS • 31

orbit north of Cho Do Island for pilot rescue as the war pushed farther north. Scores of high risk missions were accomplished, many of which involved night landings in unchartered waters, and long distance taxis in rough seas with aircraft so overloaded that takeoffs were ruled out. One of the most daring combat rescue missions was accomplished 60 miles behind enemy lines by Captain John Najarian when he landed his Albatross in the Taedong River, picked up a downed pilot, and took off under withering enemy fire. In another incident, an Albatross was jumped by an enemy MiG just after it rescued a downed flyer from the ocean. After unsuccessful attempts to shake the MiG, the SA-16A pilot took his aircraft down to 100 feet above the water. The MiG pilot banked into a steep firing pass for the kill, and to the crew's amazement, dove into the sea directly under the Albatross. The amphibian and its crew reportedly received credit for the first MiG kill by an unarmed air rescue aircraft.

The Albatross proved its worth in Korea, not only as a rescue craft but a versatile platform used for helicopter escort, troop transport, and medical evacuation. In the special operations role SA-16As were used by the 581st Air Resupply and Communications Wing to insert and pick up special units operating behind enemy lines. By war's end the Albatross was credited with saving nearly 900 lives from Korea's coastal waters and rivers. Among them were 81 U.S. and allied airmen, 66 of whom were rescued inside enemy territory. 3rd Air Rescue Group SA-16As, along with H-5 and H-19 helicopters, performed the bulk of rescue work, amassing a total of 16,277 sorties.

While war raged in Korea, global rescue was accomplished by more than 50 Air Rescue units worldwide. After the Korean armistice, Albatrosses flew air rescue missions from polar regions to Asia where they provided support for transport and tactical aircraft. They continued service with the 3rd ARS of the FEAF and under U.S. Air Forces Europe

Maintenance men at DaNang AB examine the blown engine of 51-7194 on November 14, 1966. Note the feathered prop and oil-stained fuselage. (Photo courtesy Tom Hansen)

(USAFE) where an air rescue network had been established since 1945. The task of air rescue in Europe fell to the 7th ARS based at Wiesbaden, West Germany, until it was relocated to Wheelus Field in Libya in 1952. Early that year the 7th was supplemented with the 9th ARS for expanded coverage of Europe. The 12th ARS was formed in late 1952 to provide more of France and Germany with air rescue protection. To broaden its air rescue capability even further, the USAFE reorganized its air rescue assets in 1953. This resulted in the formation of the 57th, 58th, 66th, 67th, 68th, 81st, 82nd, 83rd and 84th Air Rescue Squadrons. That same year the 53rd ARS was established at Keflavik, Iceland.

In September 1954 the 8th ARG was established to perform the mission of Strategic Air Command aircrew recovery in the polar region and Pacific Ocean. As part of the global rescue concept made viable during the 1950s, SA-16As provided coverage for air transport and tactical fighter units dur-

Number 51-7166 was a HU-16B assigned to the 354th Tactical Fighter Wing as a support aircraft. The colored nose stripes, from front to rear, were red, green, yellow, and blue. (Photo courtesy Norm Taylor via Dave Menard)

32 • GRUMMAN ALBATROSS: A HISTORY OF THE LEGENDARY SEAPLANE

HU-16B number 51-7194 seen here in Vietnam in February 1967, was typical of uncamouflaged Air Rescue Albatrosses which had their yellow markings removed or painted over. (Photo courtesy Terry Love)

Assigned to the Strategic Air Command, HU-16B number 51-006, wears insignia red markings and the distinctive SAC band and emblem on the rear fuselage. (Photo courtesy Miller via Terry Love)

ing the Lebanon and Formosa crises in 1958. Other early USAF units that operated the Albatross abroad included the 31st ARS at Clark AB, Philippines, the 33rd ARS at Naha Air Station, Okinawa, the 36th ARS at Tachikawa, Japan, the 55th ARS at Thule AB, Greenland, the 79th ARS at Guam, and the 95th Strategic Wing at Goose Bay, Labrador. Stateside units assigned Albatrosses included the 10th ARS at Elmendorf AFB, Alaska, the 41st at Hamilton AFB, California, the 42nd ARS at March AFB, California, the 71st ARS in Alaska, the 74th ARS at Ladd AFB, Alaska, and the 48th ARS at Eglin AFB, Florida, where Albatross training was conducted from 1964 to 1968.

In 1966 a Marine mechanic was running up the engines of an F-4 jet at DaNang when one engine went to 100% power. Unable to shut it down, he "bailed out" of the aircraft which careened down the flight line. The berserk Phantom slowed down after slicing through the nose of this Albatross enabling a pilot to jump aboard. Someone added a humorous touch to the event with a sign reserved for Albatrosses on rescue alert. (Photo courtesy David Wendt)

Albatrosses were operated by the following U.S. Air Force Commands: Far East Air Force, Pacific Air Force, U.S. Air Forces Europe, Strategic Air Command, U.S. Air Force Reserve, Air National Guard, and Military Air Transport Service (later Military Airlift Command).

Two Air Force Albatrosses were involved in unusual incidents, both of which lend credence to the epithet — "Grumman Ironworks." Based upon experiences in World War II, the Central Intelligence Agency (CIA) planned the development of classified air units during the early 1950s. One such brainchild, designed for integral Air Force operations, evolved into the 580th, 581st and 582nd Air Resupply and Communication Wings (ARCW). Formed to operate in the Far East, Europe, and Africa, these clandestine units, under CIA auspices, could penetrate foreign borders to infiltrate agents and equipment. The 581st, which compiled an impressive record of covert activities using Albatrosses behind enemy lines in Korea, operated both SA-16A and surplus B-29 aircraft. The B-29s flew long range missions with heavy loads while the SA-16A's amphibious abilities made it ideal for covert insertions and pick ups. Danger became a byword as missions were usually flown at night and at extremely low altitudes to avoid radar detection. At least three B-29s were lost and one SA-16A, serial number 51-001 of the 580th ARCW, met a similar fate.

The Albatross departed the remote training site at Mountain Home AFB, Idaho, with a crew of six on January 24, 1952. The mission was a long range navigation flight to San Diego and back. Somewhere over the desolate regions of California's Death Valley, an engine failed and the Albatross could not maintain its assigned altitude of 11,000 feet. While it remains uncertain whether the crew bailed out, the pilot rode the airplane down, and, in a superior display of airmanship, stalled the Albatross onto a mountain slope of Death Valley's Towne Summit near Furnace Creek. The crew was uninjured and made their way to Furnace Creek where they were picked up by an Albatross from March AFB.

In an area rich with lore, it was said (and even written), that the entire crew, including the pilots, bailed out and the Albatross continued on its own for another twenty miles in a gradual descent, skidded across a mountain ridge, went air-

CHAPTER 6: AIR FORCE ALBATROSS • 33

HU-16B number 51-067 carries the name "BOLL WEEVIL" on the engine nacelle and wing tip float. The radome has been included in the red arctic nose markings. (Photo courtesy Hugh Muir via Terry Love)

The tail fin tip extension added during the "B" modification usually retained its fiberglass beige color. The APU exhaust is visible in the fuselage band on 51-065 seen here in December 1974. (Photo courtesy Terry Love)

borne again at 100 knots, and finally landed on a mountain slope intact! Since the rugged terrain precluded recovery of the tenacious Albatross, it became part of the Death Valley Monument and is therefore protected by the Antiquities Act. Supervisory Park Ranger Ross Hopkins describes the SA 16A's final resting place as a "hellatious piece of territory."

On February 27, 1964, an HU-16B (S/N 51-5279) of the 48th ARS departed its home base at Eglin AFB, Florida, on a mission in support of the space program. The Albatross made an open sea landing to recover a nose cone from a missile fired at Cape Canaveral. While on the water, sea conditions rapidly worsened and the aircraft was unable to take off. For two days the Albatross taxied in rough seas toward land, consuming nearly all of its fuel. When the heavy seas subsided, there was insufficient fuel for takeoff and the flight to land, still nearly 400 miles away. Since no vessels in the area had aviation gas, the Air Force asked the Coast Guard for help. Someone suggested a fuel truck, which was loaded with aviation gas and hoisted aboard the Coast Guard Cutter HOLLYHOCK, a 179 foot buoy tender. An Air Force Colonel was taken aboard before the ship left the docks. HOLLYHOCK reached the HU-16B two days later and took it in tow. With a helicopter covering the operation, a long fuel line was floated to the Albatross from the fuel truck chained to HOLLYHOCK's foredeck. Though topped off with fuel, the Albatross and her weary crew were forced to spend another night on the water

After losing an engine during open sea practice in Drake's Bay near San Francisco in 1967, number 51-7211 was towed ashore with a bulldozer. The HU-16B of the 304th ARRS had been in the water for more than two days. (Photo courtesy David Wendt)

as the seas churned with eight to ten foot waves. The next day, the Air Force Colonel, plus supplies and JATO bottle igniters, were loaded aboard a raft and trailed astern, through wind-driven swells, to the aircraft. The pilot used the engines to close the distance and maneuver the Albatross. Even though primary swells reached six to eight feet, the Coast Guard cited the transfer as "uneventful," although this did not reflect the personal observations of the Air Force Colonel in the raft. An earlier attempt to float food and water to the Albatross crew had been unsuccessful.

HU-16B 51-029, believed to be an ANG aircraft when this photo was taken in 1967, wears a minimum of markings. (Photo courtesy Nick Williams)

Fully "cocooned" and complete with engines and drop tanks, 51-7143, a veteran with the Vietnam-based 37th ARRS, sits at MASDC in February 1975. (Photo courtesy P. Bergagnini via Terry Love)

34 • GRUMMAN ALBATROSS: A HISTORY OF THE LEGENDARY SEAPLANE

A 'wide angle' view from the cabin door while on the water in foggy Drake's Bay, showing the sea rescue platform and raft. (Photo courtesy David Wendt)

Albatross pilot Captain David Wendt in Vietnam 1965. Near Wendt's foot is the combustion heater exhaust. (Photo courtesy David Wendt)

The weather worsened and it was decided to attempt a takeoff as soon as possible. The Albatross taxied clear of the HOLLYHOCK while the fuel truck dumped oil over the side in a futile attempt to calm the seas. The first takeoff run was aborted when the starboard engine inadvertently feathered. A second attempt met with similar results, forcing a delay while the prop blade power panel was dried out. Meanwhile, the weather continued to deteriorate – tension mounted. The HOLLYHOCK cleverly took up a position on the aircraft's windward side, acting as a wind break, while following the takeoff run at full speed. It was now or never – the Albatross crew applied power, fired the JATO bottles, bounced the airplane three times and was finally airborne, to everyone's relief.

In a refueling operation not found in the manuals, the Albatross and her crew exhibited a great deal of stamina by spending five days in open seas. With an affinity for adventure, that same Albatross went on to serve as one of the few HU-16Bs that formed the 37th ARRS in Vietnam.

A similar incident in 1951 demonstrated the amphibian's adaptability to "sea legs" when it remained bound to stormy seas for two days. The SA-16A (S/N 49-082) from Flight A, 3rd Rescue Squadron, left Johnson Air Base, Japan, to rendezvous with the Navy supply ship WHITESIDE, some 400 miles out to sea. Aboard was a seriously ill sailor who would die within hours without a doctor and medical supplies. Faced with a decision that weighed the life of one man against that

Left: Captain David Westenbarger points to the latest "save scoreboard" addition, the Huggins mission in November 1965. The darker area behind the crewmen reveals where the fuselage band was painted over. Right: Some Vietnam Albatross crews kept a "save scoreboard" near the aircraft's cabin door. This one includes a HH-43 "Pedro" rescue helicopter silhouette surreptitiously applied by a rival rescue unit at Nakhon Phanom AB, Thailand, better known to airmen as "Naked Fanny." (Both photos courtesy of David Wendt)

CHAPTER 6: AIR FORCE ALBATROSS • 35

A rescue Albatross flies near Hon Me Islands, North Vietnam, during the mid 1960s. Note how the sea blue camouflage has been continued onto the float and drop tank. (Photo courtesy David Wendt)

Rough seas claim an Air Rescue Albatross off the coast of Puerto Rico in 1970. (Photo courtesy David Wendt)

of his crew, the pilot opted for a dangerous landing in the rough seas. In a tense and dramatic landing sure to evoke admiration from any aviator, the pilot full-stalled his Albatross which impacted onto the heaving seas, bounced off three swells, dug a wing into the water and righted itself. No leaks were found but a flap was severely damaged, ruling out any takeoff attempt.

After completing a hazardous raft transfer of a doctor, nurse, and medical supplies to the WHITESIDE, the ship took the Albatross in tow. For two days the aircraft endured a merciless beating from the stormy seas, being violently tossed completely out of the water only to be fully submerged by the next mountainous wave. Once safely inside Tokyo Bay, the pilot tried the engines and was astonished to hear them roar to life. He lowered the gear and taxied the battered Albatross onto land, almost 60 hours from the time the mission began. A few days later, number 9082 was back in service as though nothing had happened.

With the rapid increase of combat activity in Southeast Asia during the early 1960s, the Albatross became an early participant in the Vietnam war. In accordance with a Joint Chiefs of Staff directive in May 1964, the Air Force deployed SAR units to Southeast Asia. Two Okinawa-based HU-16Bs were sent to Korat Royal Thai Air Force Base (RTAFB) for airborne rescue control. Rigged with communications gear for the command and control role, these HU-16Bs (using call signs "Tacky 44 and 45"), worked in concert with Kaman HH-43 "Huskie" rescue helicopters and Douglas A-1 "Skyraider"

Finished in overall light gray with arctic markings, HU-16B number 51-7142 awaits its fate at MASDC in December 1974. (Photo courtesy Terry Love)

36 • GRUMMAN ALBATROSS: A HISTORY OF THE LEGENDARY SEAPLANE

Above: Wearing fluorescent red-orange trim, HU-16B 51-5295 has its engine nacelle panels removed for maintenance. This Albatross later went to the Spanish Air Force. (Photo courtesy T. Cress via Dave Menard)

escort planes to form a rescue task force. With a limited electronic search capability, the Albatrosses orbited above 4,000 feet (out of small arms fire range) to search for survivors. Over water the Albatross could also do a visual search and double as the rescue vehicle. In the command and control role, the Albatross was, in effect, an early version of the renowned Forward Air Controller (FAC), tasked with locating downed airmen, radio relay, and coordinating and directing rescue missions.

As the war over Laos and North Vietnam intensified in 1965, it became increasingly difficult for the HU-16, having limited communication gear and meager crew comforts for such missions, to continue rescue task force control. They

51-7194 shares a crowded ramp at DaNang AB, Vietnam, in late 1966. (Photo courtesy Tom Hansen)

With most of its ordnance racks empty, a RESCAP Skyraider pulls alongside the Albatross following the Huggins mission on November 1, 1965. Headed back to DaNang, the HU-16B's bomb rack is empty after the fuel tank finally dropped off. (Photo courtesy David Wendt)

Douglas A-1 Skyraiders served as armed escorts for rescue Albatrosses in Vietnam. Known as RESCAP, a trio of Navy A-1s fly off an Albatross' wing over Vietnam. (Photo courtesy David Wendt)

CHAPTER 6: AIR FORCE ALBATROSS • 37

SA-16A 51-7184 in July 1955. Note the JATO bottle and absence of de-icer boots. This Albatross was later upgraded to a "B" Model and turned over to the Philippine Air Force. (Photo courtesy W.J. Balogh via Dave Menard)

were replaced by Douglas SC-54 "Rescuemasters" which could stay aloft longer, fly at higher altitude, provide more room for the crew, and carry the latest electronic and communication equipment.

Concurrent with the pair of HU-16s sent to Korat, the 31st ARS at Clark Air Base in the Philippines sent three HU-16Bs to DaNang Air Base, South Vietnam, for aircrew rescue missions in the Gulf of Tonkin off the coast of North Vietnam. An additional two were sent in July to meet increased demands placed upon the Air Rescue Service by a rapid buildup of forces. The two Thailand-based HU-16s, which had been repositioned further north at Udorn RTAFB, also joined the Albatross contingent at DaNang. These aircraft formed the 37th Aerospace Rescue and Recovery Squadron (ARRS) which flew long duration standby missions over the Gulf. The 37th was one of four squadrons dispersed throughout Southeast Asia whose parent unit was the 3rd ARRG of the 7th Air Force which was responsible for all SAR efforts in the theatre.

The routine Albatross flights were often punctuated by daring rescues close to North Vietnamese shore gun batteries and often in horrid weather conditions. If sea conditions permitted, Albatrosses made water landings to pick up survivors. If a water landing was ruled out, a rubber raft and survival equipment could be dropped. Often, a pararescueman (better known as "PJ" from the shortened parajumper designator), would jump or parachute into the sea to assist. Besides the PJ, a typical Albatross crew comprised a rescue crew commander (pilot), copilot, navigator, flight mechanic and radio operator. It was common practice for Albatross

A camouflaged HU-16B of the 33rd ARRS prepares to depart Okinawa for the splashdown of Gemini 8 in March 1966. (Photo courtesy David Wendt)

HU-16B number 51-071 is dwarfed by a Douglas C-124A "Globemaster" in 1964. Seen here in the arctic scheme, 071 was blown out of the water off North Vietnam two years later. (Photo courtesy David Wendt)

38 • GRUMMAN ALBATROSS: A HISTORY OF THE LEGENDARY SEAPLANE

SA-16A of the 74th ARS at Ladd AFB, Alaska, in May 1957. (Photo courtesy Dave Menard)

HU-16B 51-7144 of the 304th ARRS flies past Oregon's majestic Mt. Hood during the early 1960s. The fore and aft drop tank tips have been painted to match the fluorescent red-orange arctic markings. (Photo courtesy David Wendt)

All black SA-16A number 51-016 of the Rhode Island ANG in July 1957 over Cape Cod. The white rectangle below the cockpit framed the letters "BALOFIA." (Photo courtesy P. Paulsen via Dave Menard)

Gloss black HU-16B of the Maryland ANG. The unit emblem is worn on the tail fin. Note the boarding ladder in the wheel well. (Photo courtesy R.C. Seely via Dave Menard)

Number 51-7144 in its final paint scheme of overall light gray in 1973 assigned to the 301st ARRS at Homestead AFB, Florida. The HU-16B was retired that year after 21 years of service which included four paint schemes and assignment to seven units. (Photo courtesy Terry Love)

HU-16B 51-019 in service with the Rhode Island ANG. The engine nacelle was painted black while the drop tank was probably dark blue. (Photo courtesy R.C. Seely via Dave Menard)

crews from the Okinawa-based 33rd and Philippine-based 31st ARRS to fly missions with the 37th ARRS at DaNang. One such mission proved to be anything but routine.

On March 14, 1966, an F-4C "Phantom" jet (call sign "Pluto 2") was badly damaged by anti-aircraft fire during a strike against an enemy bridge in North Vietnam. Knowing the chance of rescue was greater if he reached the Gulf of Tonkin, the pilot nursed his crippled jet to the coast where he and his copilot/systems operator bailed out. Twenty minutes away to the south, Captain David Westenbarger and his Albatross crew, entering the sixth hour of their on-station orbit, heard the "mayday" call from "Pluto 3" that his wing man had been hit and bailed out. Westenbarger turned his HU-16B (S/N 51-071) toward the gulf and requested that two "Fetch" rescue helicopters from a nearby Search and Rescue destroyer be sent to the area.

The pilots also radioed an urgent request to launch two "Skyraider" fighter-bombers (call signs "Arab 506 and 510") to provide cover to the survivors. As the Albatross arrived on scene, the crew spotted the Phantom pilots who 'splashed down' about two miles form shore. Westenbarger jettisoned the external fuel tanks and full-stalled his Albatross into the

CHAPTER 6: AIR FORCE ALBATROSS • 39

This trio of Maryland ANG SA-16As, serial numbers 51-037, 51-7157, and 51-015 at Otis AFB in 1958, illustrate the interesting contrast in ANG color schemes. (Photo courtesy P. Paulsen via Dave Menard)

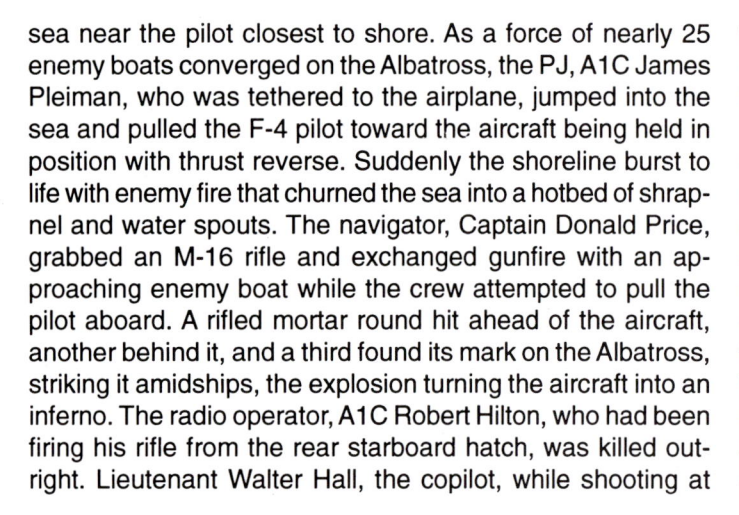

This all black SA-16A is believed to be photographed in Korea while assigned to Air Resupply and Reconnaissance Wing. The all black B-17 in the background undoubtedly served the same purpose. (Photo courtesy R. McNeil via Dave Menard)

sea near the pilot closest to shore. As a force of nearly 25 enemy boats converged on the Albatross, the PJ, A1C James Pleiman, who was tethered to the airplane, jumped into the sea and pulled the F-4 pilot toward the aircraft being held in position with thrust reverse. Suddenly the shoreline burst to life with enemy fire that churned the sea into a hotbed of shrapnel and water spouts. The navigator, Captain Donald Price, grabbed an M-16 rifle and exchanged gunfire with an approaching enemy boat while the crew attempted to pull the pilot aboard. A rifled mortar round hit ahead of the aircraft, another behind it, and a third found its mark on the Albatross, striking it amidships, the explosion turning the aircraft into an inferno. The radio operator, A1C Robert Hilton, who had been firing his rifle from the rear starboard hatch, was killed outright. Lieutenant Walter Hall, the copilot, while shooting at

the enemy boats from his overhead hatch, yelled, "Fire, we're on fire!" The flight mechanic, SSGT Clyde Jackson III, was blown out of the hatch and seriously injured.

The PJ yelled to Price from the water that he was hit and needed help, but his calls went unanswered since the initial explosion slammed Price against a bulkhead, causing him to nearly lose consciousness. While trying to exit the blazing, sinking Albatross, Price was burned by searing heat from igniting JATO bottles. Once in the ocean, he found the F-4 pilot and pulled him clear of the devastated HU-16, maneuvering through and under fuel burning on the surface. Captain Westenbarger clambered through his overhead hatch only seconds before flames burst through the opening. Secondary explosions ripped through the airplane which began to sink. With its tail resting on the bottom, the HU-16's nose

HU-16B 51-7144 of the 33rd ARRS at Naha, Okinawa, in late 1966. (Photo courtesy Tom Hansen)

40 • GRUMMAN ALBATROSS: A HISTORY OF THE LEGENDARY SEAPLANE

An Albatross crew is briefed under the wing of their aircraft at Naha, Okinawa, in January 1967. The crewman at far left wears a wetsuit indicating he is a "PJ" for the mission. (Photo courtesy Tom Hansen)

Number 51-7144 belches smoke as the engine is turned over prior to a mission from Okinawa in January 1967. (Photo courtesy Tom Hansen)

Line-up of Air Rescue HU-16Bs at DaNang AB in November 1966. (Photo courtesy Tom Hansen)

SA-16B 51-052 with 300 gallon drop tanks which were standard for Air Force Albatrosses. (Photo courtesy Nick Williams)

This HU-16B of the 302nd ARRS, which was later relocated to Luke AFB, Arizona, is finished in the later overall gray scheme. Number 51-047 carries a Brigadier General's placard above the cockpit while printing beneath the cockpit reads "Pilot Maj. Joe Foster." (Photo courtesy Terry Love)

CHAPTER 6: AIR FORCE ALBATROSS • 41

jutted from the water, burning furiously. Finally, the Albatross slipped beneath the surface, taking with it the mortally wounded radio operator and the PJ, still tethered to the aircraft.

The Navy H-3C helicopters, which Westenbarger had wisely summoned enroute, arrived and rescued him, his co-pilot, and flight mechanic plus both Phantom pilots. Though seriously injured himself, Price helped the injured F-4 pilot into the rescue sling. As a chopper converged on Price, exploding artillery rounds severed its fuel lines, forcing its return to ship. Exhausted and wounded, Price found himself alone in the sea. He crawled into an F-4 pilot's raft and it wasn't until it began to fill with red liquid that he realized how badly he was hit. He immediately discovered that the raft was drifting toward shore so he flopped back into the water and swam seaward, pulling the raft with him. Overhead, the Skyraiders pounded the shore batteries and approaching boats in an attempt to direct enemy attention away from the rescue. The helicopter returned but Price was unable to get into the horse collar sling but, with sheer might and determination, hung on with one hand as the chopper flew out of the area while reeling him in. Having previously been awarded two Distinguished Flying Crosses as an Albatross crewman, Price received the Air Force Cross and once again, took his rightful place among the ranks of dedicated rescuemen.

Captain Westenbarger was no stranger to the perils of flying Albatrosses in Vietnam. Nearly five months earlier, on November 1, 1965, he and Captain David Wendt were on patrol when they received a call from a pilot that his wing man had taken a direct hit and bailed out with a "good chute." Spotting the downed pilot's orange smoke marker, they took the Albatross down and found the RF-101 "Voodoo" pilot,

HU-16B number 51-7180 of the Air Force Reserve has red-tipped drop tanks. (Photo courtesy Terry Love)

Captain Norman Huggins, in the water near shore. It turned out that Huggins swam ashore only to be greeted by gunfire so he hastily returned to the water.

Bad weather was closing in as the pilots prepared for landing. Air Force policy dictated that Albatrosses shed their external fuel tanks prior to water landings but one tank hung up and refused to release; even with violent rocking of the wings and repeatedly working the release lever, it wouldn't budge. Nevertheless, Westenbarger and Wendt lined up for the touchdown. They had just cleared a heavy downpour when the stubborn fuel tank dropped off of its own accord. On descent, the aircraft received machine gun fire from an approaching sampan which Westenbarger ordered taken out by orbiting RESCAP Skyraiders. As the Albatross flared over the water for landing, Skyraider rockets slammed into the enemy boat, hurling wood and debris against the HU-16 and through the props, producing a nerve-shattering rattling din that threatened the crew's intense concentration. On landing, the Alba-

Sea blue-camouflaged HU-16B at Osan AB, Korea, in May 1968. Note the weathering on the wing tip float revealing the aluminum paint underneath. The red lettering on the drop tank reads, "Stand clear, explosive squibs installed." (Photo courtesy S.H. Miller via Dave Menard)

HU-16B number 51-7161 was a 33rd ARRS aircraft loaned to the 37th ARRS at DaNang where it is seen here on a taxi ramp. All yellow trim, including the edges of the rescue band on the tail, has been removed. (Photo courtesy Tom Hansen)

51-7199 on the DaNang ramp in December 1966. (Photo courtesy Tom Hansen)

Before it began service with the 37th ARRS in Vietnam, HU-16B 51-5279 was assigned to the 67th ARRS in Europe. (Photo courtesy Tom Hansen)

CHAPTER 6: AIR FORCE ALBATROSS • 43

The hulk of an all black Air Force SA-16A lies derelict in the desert at MASDC in 1972. (Photo courtesy Chuck Pomazal)

This Alaska-based SA-16A wore a distinctive red arrow which incorporated the cabin windows. Number 51-474 was later converted to a SHU-16B and delivered to Norway. (Photo courtesy U.S. Air Force via Dave Ostrowski)

tross crew discovered that Huggins was not alone — other heads in the water belonged to enemy swimmers Huggins had been holding off with his .38 revolver.

The pilots interposed the HU-16 and the PJ, A1C James Pleiman, swam out with a tether attached and retrieved Huggins (the practice of tethering PJs to the aircraft enabled them and their survivors to be pulled in faster and safer, however the procedure later proved disastrous for Pleiman). Meanwhile, the flight mechanic kept the enemy swimmers at bay with an M-16 rifle.

Once inside the Albatross, Huggins grabbed an M-16 to shoot at his "swim mates." He then clambered into the cockpit dripping wet and covered with yellow-green sea marker dye shouting, "Those sons-of-bitches shot me down!" Still on an adrenalin high, he had to be told repeatedly to sit down for the takeoff but finally just squatted between the pilot's seats, rising after liftoff to point out the gun site that shot him down. Flying an unarmed rescue airplane, Westenbarger and Wendt knew their part in the Huggins mission was over. They, along with Pararescueman Pleiman, exemplified the spirit of Air Rescue which earned them the Silver Star.

David Wendt logged more than 1,000 hours in 33rd ARRS Albatrosses flying over 150 combat missions. After his Vietnam tour, he continued to fly the HU-16 until it was retired in

Maintenance men pour over SA-16A number 51-7142 at Eglin AFB, Florida, during the 1950s. (Photo courtesy U.S. Air Force via Dave Ostrowski)

1972.

A second Albatross was lost over Vietnam on October 18, 1966. The HU-16 (S/N 51-7145) departed DaNang about 11:00 AM in marginal weather for the afternoon orbit over the gulf which was code named "Crown Bravo" — "Crown Alpha" denoted the morning orbit. The weather worsened and the pilot failed to make a routine radio check. When the weather

With "ARCTIC ALBATROSS" printed on the nose, the first production Albatross settles over the ocean during tests to prove that triphibian gear had no affect on water operations. (Photo courtesy U.S. Air Force via Dave Ostrowski)

HU-16B 51-7199 of the 37th ARRS at DaNang AB in January 1967. (Photo courtesy Tom Hansen)

44 • GRUMMAN ALBATROSS: A HISTORY OF THE LEGENDARY SEAPLANE

cleared, a second HU-16 searched for the missing Albatross with negative results. A massive air and sea search followed but no trace of the aircraft or crew was found. Listed as MIA over North Vietnam are the pilot, Major Ralph Angstadt; copilot, 1LT John Long; navigator, Major Inzar Rackley Jr.; flight mechanic, TSGT Robert Hill; PJ, A2C Stephen Adams; radio operator, A1C Lawrence Clark, and flight mechanic, MSGT John Shoneck. All were members of the 37th ARRS except Shoneck, an HH-43 flight mechanic from the 38th ARRS probably amassing flight hours since "Huskie" crews were hard pressed for flight time. Two additional Albatrosses were lost in the Southeast Asian theatre due to operational causes on February 9th and July 17th, 1967. Despite the small number of HU-16s operated by the ARRS for the Vietnam war effort, losses were high with a total of four aircraft and nine crewmen. Air Rescue crews and their Albatrosses did however, distinguish themselves by saving the lives of 26 U.S. Air Force and 21 U.S. Navy airmen during combat operations. The last Albatross combat mission was flown over the Gulf of Tonkin on September 30, 1967. Conceding the limits of their Alba-

Air Force crewmen nicknamed their Albatrosses "Dumbo" after Walt Disney's flying elephant. The popular character graced the 303rd ARRS emblem, which was designed in 1957 and applied to the unit's SA-16As. The Latin phrase, which means "Serving to Save," summarized the squadron's mission.

Below: SA-16A serial number 49-071 was the 29th Albatross built. Its first assignment was Flight A, 6th Rescue Squadron at Goose Bay, Labrador. It is seen here at Goose Bay Airport on September 25, 1950, after landing in Lake Winokapau to pick up a B-50 crew that parachuted into the wilderness. (Photo courtesy the Smithsonian Institution)

CHAPTER 6: AIR FORCE ALBATROSS • 45

An HU-16B settles into an open sea landing. Clearly visible are the wide painted black areas to mask the engine exhaust. (Photo courtesy David Wendt)

trosses, the Air Force retired them from the Pacific rescue inventory after more than 18 years of humanitarian and combat rescue. The venerable HU-16s were replaced by Sikorsky HH-3E helicopters which could air refuel and land on water.

As testimony to the danger of rescue work in open seas, the Air Force suffered heavy Albatross losses outside the Vietnam theatre. In June 1965, an HU-16 from the 31st ARRS landed in 14 foot swells to pull four survivors from the Pacific Ocean after two B-52s collided and crashed enroute to Southeast Asia. The survivors and two PJs were transferred to a passing Norwegian freighter, while the HU-16 crew stayed with their aircraft, which had been damaged during the rescue. Immediately after the crew was taken aboard a Navy ship, the Albatross sank.

In 1970 an HU-16B of the 301st ARRS, accompanied by another Albatross, crashed into the Atlantic Ocean and sank three miles west of Ramey AFB, Puerto Rico. The aircraft went prematurely airborne on a practice takeoff and settled back onto the water when a huge swell heaved the plane back into the air. That sheared off the right float and caused the left engine mounts to fail. The crippled Albatross rolled to the right and nosed into the water. Within 15 minutes the reigning sea claimed its quarry and all six crewmembers were res-

An HU-16B begins a touchdown in a rough sea. Landings and JATO takeoffs in rough seas humbled many an Albatross pilot. (Photo courtesy David Wendt)

SA-16A number 50-172 was the first of a production batch built in 1950. The rectangle on the bow was yellow-orange with a black border. The upper wing area between the engine nacelles is the same color with large black RESCUE letters. (Photo courtesy the Smithsonian Institution)

HU-16B 51-7211 at the final phase of a JATO takeoff. (Photo courtesy David Wendt)

The pilot of 51-5279 uses the engines to maneuver near the USS HOLLYHOCK while crewmen in the bow hatch secure tow lines. (Photo courtesy U.S. Coast Guard)

The carrier USS LAKE CHAMPLAIN maneuvers clear of the HOLLYHOCK and the stranded Albatross. (Photo courtesy U.S. Coast Guard)

cued by the second Albatross.

During its heyday, Grumman Aircraft Corporation sponsored an honorary organization called "The Albatross Club," whose purpose was to recognize the outstanding contributions of a select group of U.S. Air Force pilots who participated in worldwide rescue operations with the Albatross. Membership was open to all Albatross pilots associated with the Aerospace Rescue and Recovery Service.

The Albatross Club met annually to present awards to pilots who participated in air-sea rescues during the current year. The rolls of these participants were maintained by Grumman's Public Relations Department.

During the 1960s the war in Southeast Asia shared the news headlines with American space flights. Some Albatrosses of the 33rd ARRS at Naha Air Base, Okinawa, played an important part in the MERCURY and GEMINI space flights flying "splashdown" alerts over the Western Pacific. Committed to the effort was a mix of HC-97, SC-54, and HU-16B aircraft whose duties were search, Pararescuemen drops, radio relay, and if necessary, astronaut retrieval in open seas.

During 1954 and 1955 the Air National Guard was allowed to form four Air Resupply Squadrons in lieu of jet-equipped units that were relocated. The first unit organized was the 129th at Hayward Municipal Airport, California, in the Fall of 1954. Maryland's 135th was formed in September 1955 at Harbor Field, Baltimore, and relocated to the Glenn L. Martin Airport in 1960. The following month the 130th was established at the Kanawha County Airport, West Virginia. Last was the 143rd organized at Rhode Island's T.F. Green Airport in November 1955.

The 135th and 143rd initially received both Curtiss C-46 "Commando" transports and SA-16As, while the 129th and 130th began operations with C-46s with SA-16As added during 1956. Those squadrons were redesignated Troop Carrier Squadrons in 1958 and changed to Air Commando Squadrons in 1962 and 1963.

The 129th ceased operations with Albatrosses in 1963 and West Virginia's were replaced the following year. During

CHAPTER 6: AIR FORCE ALBATROSS • 47

A crewmember in the bow hatch disconnects the tow line from the pendant prior to a takeoff attempt in rough seas. (Photo courtesy U.S. Coast Guard)

The entire Albatross crew works to handle the fuel hose floated to the aircraft from the fuel truck lashed to HOLLYHOCK's foredeck as the aircraft heaves in rough seas. (Photo courtesy U.S. Coast Guard)

1968 those units underwent still another name change, that time to Special Operations Squadrons. The 135th and 143rd retained their Albatrosses until 1971. Both SA-16A and -16Bs (later HU-16B) were used by those units primarily for support of Special Forces.

Albatrosses served with five U.S. Air Force Reserve units which were dispersed throughout the country. Those Air Rescue Squadrons (redesignated Aerospace Rescue and Recovery Squadrons in the mid 1960s) operated SA-16As and HU-16Bs from their inception. They were the 301st at Homestead AFB Florida; 302nd at Long Beach Municipal Airport California (relocated to Luke AFB, Arizona); 303rd at March AFB, California; 304th at Portland International Airport Oregon; and the 305th at Selfridge ANG Base Michigan.

When the Dominican Republic crisis flared up in 1965, Albatrosses of the 301st ARRS flew nearly 130 hours evacuating Americans from the beleaguered island. The unit converted to a composite squadron in 1971 when it added HH-34 helicopters to its inventory. By 1973 all the 301st HU-16s had been replaced by the HH-34. In Fall 1966 the 302nd received the last three HU-16Bs to serve in the Pacific from the 33rd ARRS at Okinawa. The 302nd, 303rd, and 304th retired their Albatrosses in 1971. The 303rd and 305th replaced theirs with HC-97s in 1966 which, though incapable of water landings, were faster, able to carry more equipment, and stay on station longer. For its finale as an Air Force Reserve rescue

This view of HU-16B 51-7180 shows the drooped wing leading edge and squared wing tips familiar to the "B" Model Albatross. The float tip is adaptable for a triphibian skid. (Photo courtesy the Grumman Corporation)

48 • GRUMMAN ALBATROSS: A HISTORY OF THE LEGENDARY SEAPLANE

This Albatross of the 71st ARS exhibits its triphibian capabilities on Lake Louise, Alaska, in February 1959. (Photo courtesy U.S. Air Force Air Combat Command)

The veteran seaplane tender USS CURRITUCK prepares to offload HU-16B number 51-7144 of the 37th ARRS at DaNang following an engine change on October 13, 1966. A Pacific tour begun during late 1966 had the CURRITUCK stationed off Vietnam's shore. (Photo courtesy U.S. Navy via Bob Lawson)

HU-16B number 51-5290 about to "get up on the step" in 1964 during a water takeoff. The Albatross wears a USAF Distinguished Unit Award below the cockpit. Barely visible on the wings are wide yellow-orange bands which serve as a background for USAF markings. (Photo courtesy U.S. Air Force Air Combat Command)

aircraft, an HU-16B of the 304th ARRS was instrumental in saving the life of an injured seaman aboard a vessel 600 miles off the Oregon coast. Three PJs jumped from the Albatross into the ocean and were taken aboard the PECHENGA. This was the first paradrop ever made in Pacific Northwest waters. The last Air Force Albatross (S/N 51-5282), a Reserve aircraft, was retired to the Air Force Museum, but not before it completed a world record altitude flight on July 4, 1973, in the hands of a 301st ARRS crew.

The first production Air Force Albatross was painted overall aluminum as an anti-corrosion measure which became the standard ARS/ARRS scheme for USAF SA-16As, SA-16Bs, and HU-16Bs. Standard practice had orange-yellow (also called chrome or insignia yellow) markings, edged with six inch black stripes, applied in the form of a 36 inch wrap-around band on the rear fuselage and wing tips. Floats and their struts were also orange-yellow. Machines produced from

In its sinister black weathered paint, this HU-16B evokes an image of anything but grace and beauty. Serial number 51-7169 was probably flown by an Air National Guard unit. The prop warning stripe was white. (Photo courtesy Bob Lawson)

An SA-16A prepares to alight on the water. Number 51-7161 was landing near a ship. (Photo courtesy U.S. Air Force Air Combat Command)

CHAPTER 6: AIR FORCE ALBATROSS • 49

The water-borne Albatross maneuvers with its engines to recover the tow line floated from the HOLLYHOCK. (Photo courtesy U.S. Coast Guard)

HOLLYHOCK crewmembers prepare to float a fuel hose to the Albatross. (Photo courtesy U.S. Coast Guard)

1948 to 1950 carried their serial numbers on the hull in large yellow figures, also edged with black. The upper wing area on these early examples, between the outer edges of the engine nacelles, was painted orange-yellow with a black border. Superimposed in this area was the word RESCUE in black. An orange-yellow rectangle, with the usual black trim, was on the forward fuselage (bow). This was either left blank or framed unit designations or USAF. During 1951 this was replaced with the word RESCUE. Throughout the 1950s a number of Air Force Albatrosses featured variations of the standard scheme having portions of the upper wings painted orange-yellow with black edging. The walkway on top of the fuselage was either solid black, a black outline, or orange-yellow with black edges. Black-painted exhaust areas were also common. The emblem of the Military Air Transport Service (later Military Airlift Command), the parent command, was usually worn just aft of the fuselage band. The last four digits of the serial number were carried on the nose gear doors on an orange yellow background with black edges.

The word RESCUE on the forward fuselage in the first operational scheme would later be changed to U.S. AIR FORCE in insignia blue or black. A later scheme, introduced during the early 1970s of overall medium gray, brought changes that included an insignia blue RESCUE band, trimmed with yellow, on the vertical fin.

Those aircraft assigned to arctic and Alaska-based units originally had insignia red tail surfaces, noses, and wing sections added to the basic scheme. This was later changed to

An HU-16B (S/N 51-5306) of the 67th ARS stands ready on the ramp at Prestwick, England, during the 1960s. (Photo courtesy MAP)

50 • GRUMMAN ALBATROSS: A HISTORY OF THE LEGENDARY SEAPLANE

fluorescent red-orange which could be applied as paint or pressure-sensitive film. Control surfaces remained the aircraft color. A number of Albatrosses flown by Air Resupply and Communication Wings were observed in overall black finish with red figures. Especially interesting was the color scheme unique to HU-16Bs of the 33rd and 37th ARRS which operated from DaNang AB, Vietnam. Intended to hide the aircraft from enemy gunners during water operations, all upper surfaces were painted dark sea blue with camouflage gray undersides. A nondescript demarcation line varied between sharp scalloped edges or a gradual blending of the two contrasting colors. Uncamouflaged HU-16Bs in Vietnam had their high visibility yellow markings painted over in silver to match the aircraft color.

In keeping with the clandestine nature of their special operations support missions, most Air National Guard Albatrosses were finished in overall gloss black with red figures. A number of ANG HU-16Bs also appeared in overall dull silver marked only with national insignia and serial numbers. Albatrosses of the Air Force Reserve retained the basic ARRS scheme having the USAFR emblem usually positioned above the RESCUE band on the tail fin. Reserve units that inherited HU-16Bs flown in Vietnam retained the sea blue camouflage while others, including the 304th and 305th ARRS, painted some aircraft with a slightly modified sea blue camouflage.

U.S. AIR FORCE ALBATROSS PRODUCTION

48-588 to 48-607
49-069 to 49-100
50-172 to 50-182
51-001 to 51-071
51-471 to 51-476
51-5277 to 51-5306
51-7140 to 51-7255
51-15270 and 51-15271
52-136 and 52-137

The 1964 film "Flight From Ashiya" dramatized the heroic exploits of the Air Force Air Rescue Service. The story centered around an Ashiya-based Albatross squadron. (Photo by author)

SA-16A S/N 51-035 of Maryland's 135th Air Commando Squadron wears the unit's emblem on the tail during 1963. (Photo courtesy Steve Miller)

SA-16A of the same unit wears a different set of markings during the same time period. (Photo courtesy Steve Miller)

A California ANG HU-16B sports its 129th Troop Carrier Squadron emblem on the aft fuselage in 1963. (Photo courtesy MAP)

CHAPTER 6: AIR FORCE ALBATROSS • 51

CHAPTER 7

COAST GUARD GOATS

The U.S. Coast Guard recognized the search and rescue potential of the Albatross during the airplane's infancy and entered the joint Air Force/Navy procurement agreement in 1950. The new amphibians were slated as replacements for the mix of aircraft flown by the Coast Guard since World War II. Those entering Coast Guard service were designated UF-1Gs and paralleled the specifications for the Navy UF-1. The first example was accepted at the Grumman plant and flown to USCG Air Station Brooklyn, New York on May 7, 1951. The initial production batch of four aircraft was assigned USCG numbers 1240 to 1243. Following that was an order for nine aircraft numbered 1259 to 1267, the first of which was delivered in March 1952. From March 1953 to January 1954 a group of ten UF-1s, numbered 1271 to 1280, was commissioned.

After the Korean war a number of Coast Guard Albatrosses were re-purchased by the Navy whose air assets were strained by the conflict. The sale involved the first five aircraft from a group of seven UF-1Gs (1288 to 1294) delivered to the Coast Guard during April 1954. Numbers 1288 to 1292 were added to the Navy inventory between December 19, 1954 and January 11, 1955. About the time the Korean Armistice was signed, the Coast Guard became wholly responsible for search and rescue over U.S. waters including those of the U.S. Trust Territories. To fulfill this commitment, fifteen SA-16As ordered by the Air Force were transferred to the Coast Guard prior to completion and numbered 2121 to 2135. This segment of the Albatross fleet expansion was accomplished by February 1954. Beginning in February 1956, 40 SA-16As in service with the Air Force were transferred to the Coast Guard. Their four-digit USCG number was derived by deleting the first number(s) and hyphen of their Air Force serials. On August 14, 1955, the second XJR2F-1 prototype was transferred from the Navy to the Coast Guard Air Station at Elizabeth City where it served as a training aid.

By January 1957, Grumman had produced 36 UF-1G Albatrosses for the Coast Guard. Commensurate with Navy and Air Force programs during the late 1950s, the Coast Guard upgraded all but five of their Albatrosses to UF-2G standards. The five not improved were those resold to the Navy as UF-1s. The Coast Guard soon recovered the loss of these five aircraft during late 1956 when five units were transferred form the Navy as UF-1s (141284 to 141288), converted to UF-2Gs and serialed 1313 to 1317. A number of discrepancies are noted regarding the five UF-1Gs not brought up to UF-2G standards and, for that matter, the total number of Albatrosses in USCG service. In cross referencing available documentation, including Grumman and Coast Guard records, it becomes apparent that the five aircraft were those sold to the Navy.

The UF-2G modernization program encompassed Grumman Design Numbers G-234, G-270 and G-288, differ-

Though seen here in storage in April 1976, number 7246 still bears evidence of its use as a test platform for the AOSS system. (Photo courtesy Nick Williams)

HU-16E number 1311 was built as a single production unit. Here it awaits its fate at the desert storage yard at Davis Monthan AFB, Arizona. (Photo courtesy Nick Williams)

HU-16E number 1311 at MASDC in April 1976. (Photo courtesy Nick Williams)

entiated only by minor systems installations. After modification to UF-2G standards, an aircraft from the second production batch (USCG No. 1261), underwent flight test evaluation by the U.S. Navy's NATC at Key West, Florida. The purpose of the trials, which were monitored by the U.S. Air Force, was to evaluate the UF-2G's flying qualities at high gross weights. The evaluation period began on November 5, 1958, and ended on January 15, 1959, with favorable recommendations for water operations at gross weights up to 32,000 pounds. All Coast Guard UF-2Gs were redesignated HU-16Es in 1962. Sometime during its career the Albatross was affectionately dubbed the "Goat" by Coast Guard crews. While the nickname's origin remains uncertain, it was used with respect and sarcastically expanded to "Whispering Goat" in reference to the Albatross' deafening roar during takeoff.

In Coast Guard service the Albatross performed a wide variety of duties from air stations at Mobile, Alabama; Annette Island and Kodiak, Alaska; San Diego, San Francisco, and Sacramento, California; Miami and St. Petersburg, Florida; Barbers Point, Hawaii; Cape Cod, Massachusetts; Traverse City, Michigan; Biloxi, Mississippi; Bermuda; Brooklyn, New York; Elizabeth City, North Carolina; Salem, Oregon; Corpus Christi, Texas; San Juan, Puerto Rico; Port Angeles, Washington, and Sangley Point, Philippines. Though rarely publi-

UF-2G number 7255 sets down in the water after setting a distance record for amphibians, flying non-stop from Kodiak, Alaska, to Pensacola, Florida. (Photo courtesy the Grumman Corporation via Hal Andrews)

UF-2G 7250 on a search and rescue mission near Honolulu during late April 1961. (Photo courtesy U.S. Coast Guard via Hal Andrews)

CHAPTER 7: COAST GUARD GOATS • 53

The second Coast Guard Albatross, number 1241, undergoes maintenance in 1966. Two years later the aircraft was placed in storage. (Photo courtesy U.S. Coast Guard via George Krietemeyer)

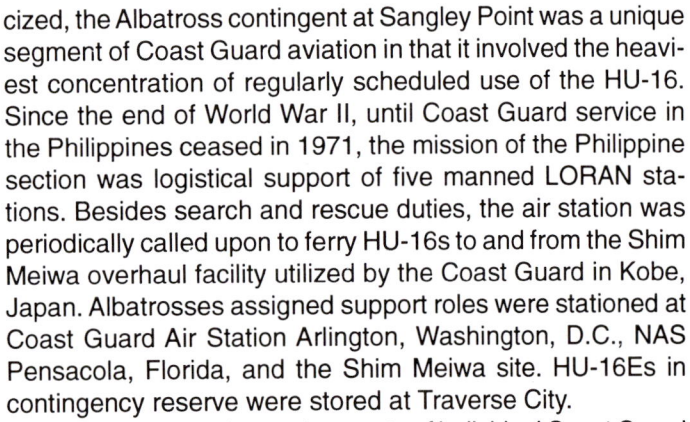

In their twilight Coast Guard years, a pair of HU-16Es on line at Miami. Numbers 7216 and 7218 were Albatrosses transferred from Air Force stocks. (Photo courtesy U.S. Coast Guard via George Krietemeyer)

cized, the Albatross contingent at Sangley Point was a unique segment of Coast Guard aviation in that it involved the heaviest concentration of regularly scheduled use of the HU-16. Since the end of World War II, until Coast Guard service in the Philippines ceased in 1971, the mission of the Philippine section was logistical support of five manned LORAN stations. Besides search and rescue duties, the air station was periodically called upon to ferry HU-16s to and from the Shim Meiwa overhaul facility utilized by the Coast Guard in Kobe, Japan. Albatrosses assigned support roles were stationed at Coast Guard Air Station Arlington, Washington, D.C., NAS Pensacola, Florida, and the Shim Meiwa site. HU-16Es in contingency reserve were stored at Traverse City.

Pinpointing station assignments of individual Coast Guard Albatrosses can prove to be an arduous task since aircraft sent in for overhaul were replaced by newer ones to maintain an air station's availability rate. That rotational system made it possible for aircraft to be assigned to numerous stations throughout their career. Overhauls were accomplished at the Coast Guard's Aircraft Repair and Supply Center (ARSC) at Elizabeth City. The rework program itself was established in 1954, reaching full swing by 1958.

The second Albatross acquired from the Air Force, number 1016 after its arrival at MASDC in May 1979. (Photo courtesy Terry Love)

Search and rescue was the Albatross' forte and as such, all USCG variants were eventually standardized with improved electronic systems to enhance their SAR proficiency. Included in those avionics packages were HF SSB transceivers, MF/VHF/UHF direction-finding gear and interrogators. That equipment would prove vital in the HU-16E's twilight when it flew

HU-16E 7213 at Cape Cod in October 1982. (Photo courtesy Terry Love)

An Albatross crew scrambles to their aircraft in July 1967. (Photo courtesy U.S. Coast Guard via George Krietemeyer)

Albatrosses assigned to Coast Guard Air Station Mobile, Alabama, carried a large yellow number on the tail fin. Number 7223 one month after its arrival at the desert storage facility in November 1972. (Photo courtesy Terry Love)

HU-16E 1026 started out as an Air Force SA-16A with serial number 51-026. Here it takes its place among the many Albatrosses placed in desert storage in November 1979. (Photo courtesy Terry Love)

SAR command and control missions. Tales abound recounting the ruggedness and perseverance of Coast Guard Albatrosses on perilous missions. Operations ranged from low altitude flights through steep winding fjords in Alaska, to massive searches in the Gulf Stream. SAR missions often involved a number of Albatrosses covering hundreds of miles of open sea for days until all chances of survivor rescue were expended. During such missions, two observers were added to the crew which otherwise matched those of Navy and Air Force Albatrosses.

Coast Guard use of the Albatross peaked during the 1960s when 71 aircraft were operational at 19 air stations in the U.S. and abroad. During that period, the heaviest concentration of missions was flown by ten aircraft from Miami's Biscayne Bay base. Small boats often broke down in the Gulf Stream prompting large scale searches. During 1972 alone, HU-16Es conducted searches for nearly 400 boats and aircraft, some as distant as 1,000 miles from home port. Albatrosses were often called upon to evacuate injured divers who were drawn to the warm beautiful coral reefs of subtropical regions. Suffering from the bends, the divers were flown to U.S. Navy decompression chambers at Ft. Lauderdale and Key West, Florida. USCG Albatrosses also flew medical evacuation missions from the many remote islands to mainland hospitals. In addition to periodic escorts of aircraft in trouble over the ocean, "Goats" were utilized as utility aircraft and VIP transports for various government agencies. During the Cuban boat exodus in the 1960s and again in 1980, Coast Guard Albatrosses flew daily SAR patrols over the waters between Cuba and Florida. It was not unusual for Coast Guard crews to find hordes of refugees fleeing Castro's Cuba, clinging to virtually anything that would float. During these operations, the Albatrosses served a dual role as airborne command posts. Cuban gunboats also plied the Florida Straits in search of the refugees. Albatross crews, armed only with sidearms, often worked at low altitude in proximity to the heavily armed boats.

Due to the age of their airframes, HU-16Es were restricted from water landings by the Coast Guard in 1972 however, they continued to prove their worth. The amphibians were often called upon as airborne command posts controlling large scale SAR operations involving ships and aircraft. This ability

The unidentified HU-16E in the foreground is fully cocooned for storage, while number 7226 to its left, in later markings, is not. (Photo courtesy Terry Love)

In a scene repeated countless times, a USCG Albatross (no. 7239) makes a JATO takeoff after picking up the ill skipper of a fishing trawler in November 1966. (Photo courtesy U.S. Coast Guard via George Krietemeyer)

CHAPTER 7: COAST GUARD GOATS • 55

A HU-16E makes a low pass at Floyd Bennett Field, Brooklyn, New York on August 27, 1966 for a flare drop at an airshow. (Photo courtesy U.S. Coast Guard via Hal Andrews)

Left: A Coast Guard crewman checks over his Albatross prior to a law enforcement patrol from Cape Cod Air Station in July 1979. (Photo courtesy U.S. Coast Guard) Right: Two Miami-based UF-1Gs demonstrate the passing of the age when dull inconspicuous colors was secondary to the safety of high visibility. The Albatross in the foreground wears an experimental scheme of overall white with fluorescent orange bands trimmed with black. White was also found useful in reflecting solar heat. Photographed on October 27, 1958, this view shows the relocation of the USCG emblem from the bow to mid fuselage. (Photo courtesy U.S. Coast Guard)

UF-1G number 1261 seen here over San Juan, Puerto Rico, served as the UF-2G prototype. (Photo courtesy U.S. Coast Guard via Hal Andrews)

The first Coast Guard Albatross, UF-1G number 1240, on display on Armed Forces Day, May 19, 1951, at Bolling AFB, Washington, D.C. (Photo courtesy U.S. Coast Guard via Hal Andrews)

A Douglas R5D-3, with an experimental high visibility paint scheme, flies off the wing of a UF-2G carrying a 1,000 pound thermate bomb in July 1953. Both aircraft flew from Argentia, Newfoundland, as part of the International Ice Patrol's Iceberg Destruction Tests. (Photo courtesy U.S. Coast Guard)

was underscored in December 1972 when a USCG HU-16E directed rescue operations following an airliner crash into the forbidding Florida Everglades. Though the advent of the helicopter joined with water landing restrictions to diminish the SAR role of the HU-16E, it was tasked with air drops in open seas. Emergency supplies such as food, water, radios, and rafts were dropped to survivors. Using the same method, pumps were supplied to leaking boats far out to sea.

The traditional SAR role of the Albatross was expanded during the 1970s to include law enforcement, specifically the interdiction of illegal drugs. A number of HU-16Es were fitted with the updated electronic equipment necessary to accomplish smuggler patrols over the Atlantic Ocean. With thousands of miles of coastline, hundreds of airports, remote airstrips, and its proximity to South and Central American countries, Southern Florida had become the threshold for the majority of drugs illegally entering the U.S. While flying these patrols, Albatrosses located numerous drug-laden vessels and prevented their cargoes from reaching the U.S. market.

Though the period marked the beginning of the end of the Albatross, several Coast Guard HU-16Es were pressed into service for environmental protection duties. Intended to discourage ships from discharging pollutants into the water, these aircraft were configured with infrared and ultraviolet gear to detect and photograph oil slicks in harbor and shipping lanes. In 1975, HU-16E number 7246 was modified to test the Airborne Oil Spill Surveillance (AOSS) system which comprised a side-looking airborne radar (SLAR) housed in a fairing at the lower rear starboard hull. Other pollution detection gear was located in the bow, in a pod attached to the lower portside fuselage, and in a modified external fuel tank beneath the starboard wing. The SLAR was the most sensitive component of the system, able to scan miles of ocean and locate an oil slick, while the other equipment could evaluate its area and density. The extensive modifications called for the wing floats and their supports to be removed. Flight tests were conducted at Edwards AFB, California, followed by assignment to Miami Air Station and later San Francisco Air Station. The AOSS system, manufactured by Seaveyor, was also helpful in mapping the area and density of ice floes as well as evaluating the size of fishing fleets. Evaluation of the AOSS led to the AIREYE system in use today aboard Coast Guard HU-25 "Guardians."

Still another unusual adaptation of the Albatross put it to use as an "Iceberg bomber." During 1958 and 1959, the International Ice Patrol (IIP) was involved with iceberg destruc-

HU-16E number 7246 served as the test platform for the AOSS system. Besides modifications to the bow for the installation of electronic gear, a SLAR pod was added to the rear fuselage, a large window replaced the starboard cabin hatch, and the wing tip floats were removed. The Seaveyor emblem is just aft of the plane number. (Photo courtesy Peter B. Lewis via Bob Lawson)

CHAPTER 7: COAST GUARD GOATS • 57

HU-16E number 1016 awaiting storage at MASDC in May 1967. (Photo courtesy Terry Love)

Under threatening skies, number 7218 sits on the ramp at Brownsville, Texas, in 1969. The drop tank carries unusual markings in the form of a black section and red arrows. (Photo courtesy Terry Love)

tion tests. As part of a research project to determine heat deterioration effects, Coast Guard UF-2Gs operating from Argentina, Newfoundland, were assigned to the IIP to drop heat bombs on icebergs.

At the height of the heaviest ice season in 1959, the Albatrosses dropped 20 bombs on selected icebergs in the Grand Banks region of Newfoundland. The 1,000 pound bombs encased a cluster of small bomblets capable of burning at temperatures of 4,300 degrees fahrenheit.

In addition to search and rescue and the wide variety of operations flown, the Coast Guard Albatross set a number of world class records for amphibious aircraft. The countless accomplishments of the Albatross during its Coast Guard career were tempered by tragedies which claimed the lives of 33 airmen and resulted in the loss of eleven aircraft. Nearly one third of those losses occurred during 1967 alone, underscoring the ominous nature of rescue work. A portion of that chronology is as follows:

On December 14, 1954, UF-1G number 2121 was flown from Annette Air Station, Alaska, to Haines, Alaska, on a medical evacuation mission. The Albatross crashed during a water takeoff, possibly due to a layer of ice which had accumulated on its wings while waiting for the patient to be deliv-

Number 7213 carries red drop tanks which were more commonly painted white in Coast Guard service. (Photo courtesy Terry Love)

58 • GRUMMAN ALBATROSS: A HISTORY OF THE LEGENDARY SEAPLANE

HU-16E 7214 tied to the ramp at Miami CGAS. (Photo courtesy Candid Aero Files)

HU-16E number 7226, now under private ownership, at San Francisco. (Photo courtesy Terry Love)

ered. The aircraft was lost at sea and AL1 C.E. Habecker, AD1 A.P. Turnier and AL3 D.E. John perished.

UF-1G number 1278 was performing a JATO demonstration at the Salem Air Station on May 18, 1957, when it stalled and crashed. LCDR A.P. Hartt, Jr. and AO2 W.J. Tarker, Jr. were killed and the aircraft damaged beyond repair.

While flying a maintenance test flight at the Brooklyn Air Station on August 22, 1959, UF-1G number 1259 crashed on its takeoff run. The crash, caused by either jammed or reversed ailerons, resulted in the deaths of LCDR C.S. Lebaw, LT R.A. Faucher, AD3 M.R. Ross and AL3 G.R. Fox.

On July 4, 1964, HU-16E 7233 was attempting an instrument approach to Annette Air Station after a long duration search mission when it crashed into a mountain. LCDR J.N. Androssy, LT R.A. Perchard, AO1 H.W. Olson, AM2 D.G. Malena and AT3 E.D. Krajniak lost their lives.

While flying a night search over the ocean in minimal visibility on March 5, 1967, HU-16E number 1240 from St. Petersburg Air Station, was lost with all hands. The Albatross crew located the vessel in distress and was preparing to drop pump equipment when it crashed into the Gulf of Mexico. Members of the crew were LT C.E. Hanna, LTJG C.F. Shaw, AD1 R.H. Studstill, AT1 C.M. Powlus, AT2 J.B. Thompson and AE3 A.L. Wilson.

HU-16E number 7237 departed Annette Air Station on June 15, 1967, in search of a downed aircraft over mountainous terrain. The Albatross flew into a box canyon and crashed

HU-16E number 1313, formerly assigned to mobile as indicated by the yellow fin number, rests pensively at MASDC in April 1972. Note how the tail fin color scheme was continued onto the trim tab. (Photo courtesy Chuck Pomazal)

Coast Guard Albatrosses share their desert roost with black-painted Air Force HU-16s in early 1972. (Photo courtesy Chuck Pomazal)

HU-16E wears the color scheme adopted by the Coast Guard after 1960. Albatrosses assigned to Barbers Point Air Station had a unit emblem applied to the tail fin as seen here in 1967. (Photo courtesy Nick Williams)

Number 1313 rests on her keel in the desert. The cabin window aft of the wheel well was usually replaced with an observation "bubble". (Photo courtesy Chuck Pomazal)

CHAPTER 7: COAST GUARD GOATS • 59

killing LT R.D. Brown, LT D.J. Bain and AT1 R.W. Striff.

Based at San Francisco Air Station, HU-16E number 2128 was conducting a search for an overdue boat along the California coastline on August 7, 1967. The boat was located and while working at an accurate fix of the vessel in low visibility, the Albatross crashed into mountainous terrain. LTJG F.T. Charles, AD3 W.G. Prowitt, and AD3 J.G. Medek perished in the incident.

HU-16E number 1271 was attempting a landing in severe weather at St. Paul Island, Alaska, on December 8, 1967, when it crashed. One crew member, AT2 F.R. Edmunds, was killed.

On September 7, 1973, HU-16E number 2123 departed Corpus Christi Air Station at night to conduct an over-water search with a Coast Guard helicopter. MK 45 flares were being dropped from the Albatross to provide illumination for the chopper when one ignited in the aircraft. The aircraft filled with smoke causing it to crash into the Gulf of Mexico. The mishap claimed Coast Guardsmen LTJG J.M. Mack, AD1 H.G. Brown, AM2 B.R. Gaskins, AT2 J.F. Harrison, and AT2 J.P. Pledger.

During 1966 the Coast Guard provided the U.S. Naval Air Development Center at Warminster, Pennsylvania, with an HU-16E for wing fatigue tests. The tests were terminated

A UF-1G exhibits the effectiveness of JATO during a water takeoff. (Photo courtesy U.S. Coast Guard)

This view of the forward cabin section of a UF-1G in 1951 shows a partial litter arrangement below which are stored "Mae Wests", parachute harnesses, an exposure suit container, and a "Gibson Girl" emergency radio transmitter. (Photo courtesy U.S. Coast Guard)

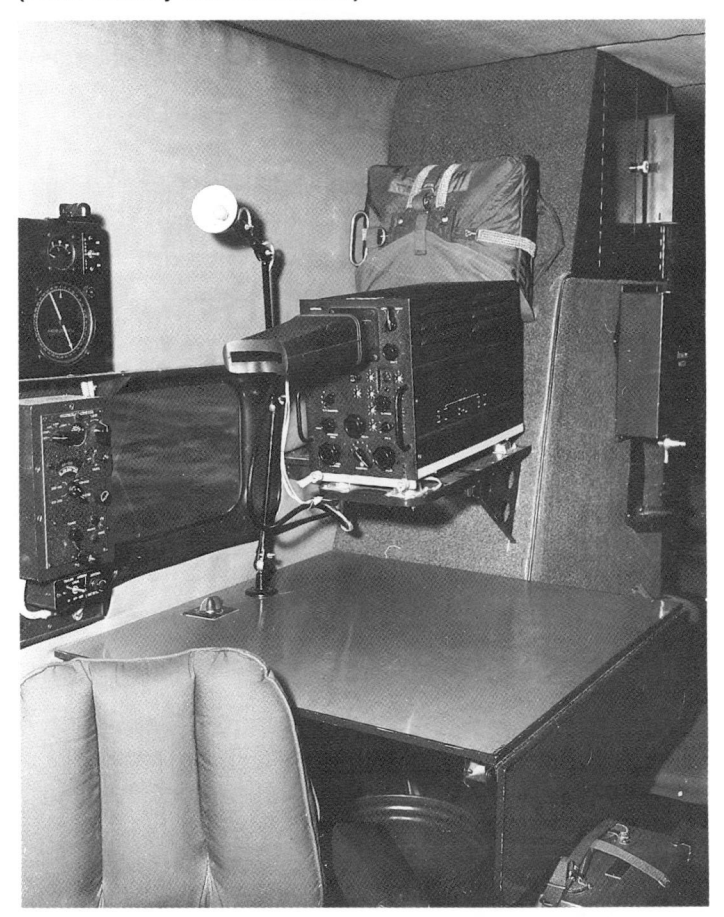

This view shows the navigator's station of a UF-1G in 1951. Pictured here is the LORAN installation mounted to the interior wheel well wall. The navigator's seat swiveled between this table and an electronic equipment console. (Photo courtesy U.S. Coast Guard)

UF-2G number 7255 during its record breaking long distance flight in 1962. (Photo courtesy National Museum of Naval Aviation)

UF-2G number 7255 in 1962. (Photo courtesy National Museum of Naval Aviation)

This UF-2G, number 7240, flies low over rough seas off the Newfoundland coast in 1959. Visible under the Albatross' wings are 1,000 pound thermate bombs for iceberg destruction tests. (Photo courtesy U.S. Coast Guard)

in October 1968 after pinpointing potential wing failure if airframes approached 19,000 hours. This resulted in a mandatory 11,000 hour retirement age for the Albatross. At that time the Coast Guard had 76 HU-16Es on inventory. By the mid 1970s that number dwindled to twenty aircraft dispersed among five air stations. Faced with the imminent demise of the Albatross, the Coast Guard decided during the early 1970s to procure jet aircraft for the medium range surveillance mission. In January 1977, a contract was awarded to Falcon Jet Corporation for 41 HU-25A "Guardians" to replace the aging Albatross. As an interim measure, 17 Convair HC-131A "Samaritans" were resurrected from storage to take over the many Albatross missions.

A UF-2G comes aboard the water ramp at Bermuda in 1961. (Photo courtesy U.S. Coast Guard)

CHAPTER 7: COAST GUARD GOATS • 61

The last operational Albatross (7250) made its final landing on March 10, 1983 at Cape Cod Air Station where it had been assigned. The majority of Coast Guard "Goats" were placed in storage while some, like their Air Force counterparts, became monuments at museums and air bases. The U.S. Coast Guard boasted a fleet of 91 Albatrosses which amassed a total of more than 500,000 flying hours. As Rear Admiral Louis Zumstein epitomized in his speech at the HU-16E's farewell ceremony, "If ever an aircraft lived up to the tradition of the Coast Guard and Coast Guard aviation, it is the Albatross."

The same year the Albatross entered Coast Guard service, a memorandum was issued by the Commandant calling for USCG amphibians to be painted with aluminum lacquer. In a scheme closely resembling that applied to Air Force

UF-1G number 2134 undergoes maintenance during the late 1950s. The fluorescent orange was later replaced with the more durable international orange. (Photo courtesy U.S. Coast Guard)

This UF-1G is painted in a brilliant experimental high visibility color scheme in 1958. (Photo courtesy U.S. Coast Guard)

The colorful Brooklyn Albatross contingent during the early 1960s. (Photo courtesy U.S. Coast Guard)

62 • GRUMMAN ALBATROSS: A HISTORY OF THE LEGENDARY SEAPLANE

A UF-2G which operated from Barbers Point flies off Hawaii's shore. (Photo courtesy U.S. Coast Guard)

This HU-16E wore a large number "1" on the tail while assigned to the Coast Guard's Avaiation Technical school at Elizabeth City during Spring 1976. (Photo courtesy Steve Zink via Steve Miller)

Albatrosses, Coast Guard amphibians wore black-bordered orange-yellow markings. Colored as such were the rear fuselage band, wing tips, walkway and floats and their supports. All figures were black and the USCG emblem was positioned on the forward fuselage below the cockpit. The four digit aircraft number appeared on the tail fin and bow.

Beginning in 1958 the Coast Guard experimented with a variety of high visibility color schemes for aircraft. After testing a spectrum of color patterns using model planes and observers, the color combination selected was white and fluorescent orange (also called blaze orange). A number of schemes were evaluated with white remaining the basic overall color. Though fluorescent orange was decidedly the best contrasting color, it soon became obvious that the finish could not withstand the effects of weather. By 1960 the switch was made to international orange trim, incorporated into a stan-

HU-16E number 1265 assigned to Mobile Air Station has a yellow "1" painted on the tail fin. This view shows how the float support outer surface was painted red while the inner surface remained white. (Photo courtesy U.S. Coast Guard)

HU-16E number 7250 was the last operational Coast Guard Albatross seen here at its last station, Cape Cod. (Photo courtesy U.S. Coast Guard)

CHAPTER 7: COAST GUARD GOATS • 63

A UF-1G undergoes maintenance at the Salem Air Station in November 1951. (Photo courtesy U.S. Coast Guard)

View of the forward cabin of a UF-1G in 1951 showing the navigator's station and emergency provisions. (Photo courtesy U.S. Coast Guard)

dard scheme on basic white. This distinctive design used orange for the fuselage band, outer wings, and outboard float supports. The nose and cockpit area became orange forming a chevron design. In addition, tail surfaces were painted orange with the tail fin done in broad alternating bands of the two colors; the station name was placed in the white band. This scheme had the USCG emblem relocated to a mid fuselage position. Engine exhaust areas and the walkway were painted black. Control surfaces were void of trim colors with the exception of the rudder on some aircraft. It also became common practice to apply the station name on the inboard side of wing floats. Albatrosses assigned to the air station at Mobile had a large yellow number painted on the tail fin while Barbers Point aircraft were observed sporting a unit emblem on the tail fin.

The scheme was revised in 1967 by replacing the international orange with a special hue called Coast Guard Red 40. The familiar chevron design gave way to a wide diagonal red band in which was centered the USCG emblem. Narrow accent stripes of white and Coast Guard Blue 41 were added immediately behind the band. The national insignia, formerly positioned below the cockpit, was relocated to the tail fin. This scheme remains the standard for modern fixed-wing Coast Guard aircraft.

A pair of Miami-based HU-16Es. Both were aircraft transferred from the Air Force. (Photo courtesy U.S. Coast Guard)

64 • GRUMMAN ALBATROSS: A HISTORY OF THE LEGENDARY SEAPLANE

Number 1240 was the first Coast Guard Albatross. The UF-1G was finished in a dull aluminum scheme very similar to that applied to Air Force Albatrosses. (Photo courtesy U.S. Coast Guard)

A fine study of a UF-1G over Long Island, New York. This view highlights the similarities to Air Force Albatross markings. (Photo courtesy the Grumman Corporation via Bob Lawson)

CHAPTER 7: COAST GUARD GOATS • 65

U.S. COAST GUARD PRODUCTION AIRCRAFT

USCG No.	Constructor No.	Commissioned Date	Remarks
1240	61	5/7/51	crashed 3/5/67
1241	66	6/29/51	decommissioned 10/22/68 to MASDC
1242	71	9/21/51	
1243	73	Sept. 1951	decommissioned 8/1/74 to MASDC
1259	106	3/25/52	crashed 8/22/57
1260	116	Jan. 1952	
1261	125	Feb. 1952	UF-2G prototype
1262	134	Mar. 1952	
1263	140	Apr. 1952	
1264	158	May 1952	
1265	160	June 1952	
1266	170	July 1952	
1267	180	9/20/52	to Philippine Air Force
1271	247	Mar. 1953	crashed 12/8/67
1272	259	Apr. 1953	
1273	271	May 1953	
1274	317	Sept. 1953	
1275	323	10/13/53	decommissioned 3/9/73 to MASDC
1276	328	11/7/53	5/13/71 training aid Elizabeth City
1277	275	May 1953	
1278	287	7/9/53	crashed 5/18/57
1279	299	July 1953	
1280	302	Aug. 1953	
1288	365	4/2/54	to USN as UF-1G BuNo. 142358
1289	366	4/9/54	to USN as UF-1G BuNo. 142359
1290	367	4/21/54	to USN as UF-1G BuNo. 142360
1291	368	4/21/54	to USN as UF-1G BuNo. 142361
1292	369	4/21/54	to USN as UF-1G BuNo. 142362
1293	370	Apr. 1954	
1294	371	5/7/54	decommissioned 6/24/73 to MASDC
1311	409	Nov. 1955	
1313	427	Aug. 1956	
1314	429	Oct. 1956	
1315	431	11/13/56	5/1/75 fire trainer Elizabeth City
1316	433	Nov. 1956	
1317	435	Dec. 1956	

UF-1G number 1260 was commissioned in January 1952. (Photo courtesy Smithsonian Institution)

AIRCRAFT TRANSFERRED FROM U.S. AIR FORCE

USAF Serial	USCG No.	C/N	Commissioned Date	Remarks
51-015	1015	88		
51-016	1016	89		
51-023	1023	97		
51-026	1026	100		
51-030	1030	104	4/3/56	decommissioned 8/1/74 to MASDC
51-7188	7188		2/18/56	decommissioned 3/4/73 to MASDC
51-7209	7209	282	April 1961	Luke AFB display Oct. 1978
51-7213	7213			
51-7214	7214			
51-7215	7215		2/22/61	12/10/74 avionics trainer at Elizabeth City
51-7216	7216			
51-7217	7217			
51-7218	7218			
51-7223	7223		3/13/61	decommissioned 11/3/72 to MASDC
51-7226	7226			
51-7227	7227		9/24/57	decommissioned 8/1/74 to MASDC
51-7228	7228			
51-7229	7229			
51-7230	7230		3/11/58	crashed 9/30/61
51-7231	7231			
51-7232	7232			
51-7233	7233		4/3/58	crashed 7/4/64
51-7234	7234			
51-7236	7236			
51-7237	7237		4/24/58	crashed 6/15/67
51-7238	7238		5/19/58	8/7/74 to U.S. Marine Corps
51-7239	7239		5/31/58	
51-7240	7240			
51-7241	7241		2/3/58	
51-7242	7242		6/16/58	
51-7243	7243			
51-7245	7245			
51-7246	7246		12/16/57	decommissioned 1/13/76 to MASDC
51-7247	7247			
51-7248	7248			to Philippine Air Force
51-7249	7249			to Smithsonian Institution N2495
51-7250	7250			decommissioned 3/10/83
51-7251	7251		Feb. 1959	Dyess AFB display 1982
51-7254	7254			
51-7255	7255		1/17/62	decommissioned 7/9/71
52-121	2121		12/30/53	crashed 12/14/54
52-122	2122		12/23/53	crashed 12/19/55
52-123	2123		1/5/54	crashed 9/21/73
52-124	2124		12/28/53	3/21/69 trainer at Elizabeth City
52-125	2125		1/14/54	decommissioned 6/12/73 to MASDC
52-126	2126		1/21/54	decommissioned 6/25/69 to MASDC
52-127	2127		1/18/54	
52-128	2128		1/29/54	crashed 8/7/67
52-129	2129		2/4/54	
52-130	2130		2/14/54	decommissioned 8/29/74 to MASDC
52-131	2131		2/10/54	
52-132	2132		2/8/54	decommissioned 8/1/74 to MASDC
52-133	2133		2/22/54	
52-134	2134		2/26/54	7/18/74 to U.S. Marine Corps
52-135	2135		2/18/54	7/17/74 to U.S. Marine Corps

Left: An HU-16E shortly after liftoff from the water at CGAS Miami in 1972. (Photo courtesy U.S. Coast Guard)

66 • GRUMMAN ALBATROSS: A HISTORY OF THE LEGENDARY SEAPLANE

A pair of Albatrosses sit on the ramp at a USCG air station believed to be Biloxi. (Photo courtesy U.S. Coast Guard)

THE U.S. COAST GUARD PTERODACTYLS

The Ancient order of the Pterodactyl was organized in the Spring of 1977 by a group of retired Coast Guard aviators at Long Beach, California. Membership in this fraternal organization is made up primarily of active duty and retired Coast Guard aviators. The balance of Pterodactyl members hail from other U.S. military services and foreign governments involved in USCG exchange programs, as well as associate members. The group's reference to "ancient" is derived from the Coast Guard's participation in the genesis of aviation and use of the extinct flying reptile symbolizes the unique aviation history of the Coast Guard.

The Pterodactyls, which boasts a number of veteran Albatross aircrewmen, not only promotes interest in and supports Coast Guard aviation, but fully supports the Coast Guard Aviation exhibit in the Naval Aviation Museum at NAS Pensacola.

CHAPTER 8

NAVAL AVIATION

The Albatross had its origins in the U.S. Navy which selected Grumman's Model G-64 to meet its requirement for an amphibious utility aircraft. Two prototypes, which the Navy designated XJR2F-1s, were named "Pelicans," although all subsequent versions of the type were named "Albatross." The prefix segment "JR" was used by the Navy from 1935 to 1955 to identify utility transports. Satisfied with the new amphibian, the Navy contracted for 32 PF-1As (BuNo 124292 to 124323) for armed patrol duties. PF-1A was an interim patrol designation for the Albatross which gave way to UF-1 for utility. Citing budget woes, the Navy lost interest in the PF-1A with the realization that the Albatross would be less proficient than the more powerful and heavily armed Martin PBM-1 "Marlin" for maritime patrol. The order was canceled and the aircraft were delivered to the U.S. Air Force as SA-16As.

Shortly thereafter, the Navy placed an order for six UF-1s, the first of which was the tenth production machine (BuNo 124374). It was first flown on December 14, 1949, and delivered to the Naval Air Test Center (NATC) at NAS Patuxent River, Maryland, on December 30th. Another UF-1 (BuNo 124376) went from the factory to the NATC on March 9, 1950, and together, the pair underwent extensive Production Inspection Trials. Designed to determine the acceptability of the UF-1, these tests were completed in February 1951, with only three flight handling deficiencies found. During successive production phases, tests were also conducted with UF-1 BuNo. 131891 during May 1953 to determine the effect of increased gross weights (up to 33,500 lbs.) during land take-offs; the flight characteristics at these weights were found to be acceptable. Concurrent tests were done to determine the value of the step vents in the hull; blocking the vents was found to have no effect on water landing stability.

Production of the Navy Albatross continued with more than 100 examples built. Of these, five were ordered as UF-1s (BuNo 141284 to 141288) but delivered to the U.S. Coast Guard as UF-1Gs serialed 1313 to 1317. An additional five were assigned Navy Bureau Numbers for administrative purposes but delivered under MAP to foreign governments. These were BuNo. 149822 to 149824 and 149836 and 149837.

The UF-1s were basically similar to the prototypes and differed from the SA-16As only in minor equipment installations. Initial production machines were fitted with AN/APS-31A search radar housed in a left underwing pod; this would be relocated as a nose radome on later production UF-1s.

A number of dilapidated Navy UF-1s are visible in this photo at MASDC in late 1979. (Photo courtesy Terry Love)

HU-16D Bureau Number 141269 is parked in an open-air service area at NAS Midway. (Photo courtesy Nick Williams)

HU-16D 137899 at NAS Atsugi in 1969. (Photo courtesy Nick Williams)

68 • GRUMMAN ALBATROSS: A HISTORY OF THE LEGENDARY SEAPLANE

Left: Despite the state of disrepair that Albatrosses at MASDC fell into, a number of them were resurrected by private and corporate owners and beautifully restored. Barely discernible on Bureau Number 131907 are tail fin markings denoting assignment to NAS Alameda. (Photo courtesy Terry Love) Right: HU-16C 137908 at MASDC in November 1979. (Photo courtesy Terry Love)

Accentuating Bureau Number 141270's white and engine gray scheme is the station assignment on the tail fin applied in Oriental-style figures. (Photo courtesy Nick Williams)

Seen here in May 1966, HU-16D 137901 served as personal hack for the Commander of the ASW Force of the Atlantic Fleet. The aircraft was finished in white and flat light gull gray. (Photo courtesy Terry Love)

HU-16D number 137911 was saved from certain extinction in the desert. It is currently a corporate-owned aircraft at Opa Locka Airport in Florida. (Photo courtesy Terry Love)

HU-16D Bureau Number 141261, seen here at MASDC in late 1979, was restored and is presently corporate-owned in Albuquerque, New Mexico. (Photo courtesy Terry Love)

Like the SA-16As, all UF variants were powered by Wright 1,425 h.p. R-1820-76A or -76B radial engines.

After the tri-service redesignation system in 1962, Navy UF-1s became HU-16Cs, the "H" denoting search and rescue and the "U" utility. Like the Air Force with its SA-16As, the Navy upgraded 33 of its UF-1s to UF-2 standard beginning in 1957. The first UF-2 conversion made its maiden flight on January 13, 1959, with the type placed in service on January 30th. After 1962 UF-2s were redesignated HU-16Ds. Bureau Numbers were assigned to a large number of UF-2s for administrative purposes to fulfill MAP contracts for foreign governments.

From the UF-1 production run, five aircraft (BuNo 131914 to 131918) were modified as UF-1Ts (later designated TU-16Cs) with dual controls and navigational training equipment. They were assigned to the U.S. Naval Academy at Annapolis, Maryland, for midshipmen pilot training. Another two were converted to UF-1Ls and later designated LU-16Cs (BuNo 142428 and 142429). Originally ordered by the USAF as SA-16As with serial numbers 51-7162 and 7164, these machines were fitted with triphibian gear and assigned to Antarctic Development Squadron Six (VXE-6) of the Naval Support Force. They flew in support of South Pole exploration named operation DEEP FREEZE beginning in 1955.

In Navy service the Albatross served primarily as search and rescue and utility aircraft while assigned to every Naval Air Station on the globe. Typical was NAS Jacksonville, Florida, where two HU-16s were assigned as station rescue

CHAPTER 8: NAVAL AVIATION • 69

Bureau Number 142428 was one of two Navy UF-1Ls fitted as triphibians for arctic duty. Here, number 428's keel skid is visible in the retracted position. (Photo courtesy U.S. Navy via Hal Andrews)

Finished in overall flat seaplane gray, this is one of five UF-1Ts built for Navy pilot training at Annapolis. UF-1T number 131918 is being fueled in the background. (Photo courtesy Merle Olmstead via Tom Hansen)

aircraft from 1960 to mid 1970. Although never used in squadron strength, the Navy assigned Albatrosses to Fleet Aircraft Service Squadrons (FASRON). FASRON 114 at Kodiak, Alaska, was the last squadron to fly the Albatross, relinquishing its aircraft in summer 1960. Thereafter, these Albatrosses were utilized at Naval Air Stations or airfields for utility transport and search and rescue. During late 1964, the Chief of Naval operations agreed with the U.S. Air Force and Coast Guard to implement a search and rescue training program for Navy Albatross pilots.

Navy Albatrosses also served headquarters commands while others were assigned to Naval Attaches (ALUSNA) in Columbia, Greece, Indonesia, and Norway. It was not uncommon for these aircraft to be flown by U.S. Marine Corps crews. One HU-16C (BuNo 137927) was assigned to the Naval Test Pilot School at NAS Patuxent River while others were used by the NATC at the same station. Two ASW SA-16Bs (later SHU-16Bs) were temporarily assigned to Patrol Squadron Thirty one (VP-31) at NAS North Island, California, in 1961. There they underwent operational trials prior to delivery to Norway. These aircraft wore serial numbers 51-048 and 51-050 in keeping with the policy that aircraft ordered under MAP/MDAP contract be delivered painted aircraft gray with USAF markings.

To provide support of PBM-1 "Marlin" patrol squadrons, Navy Albatrosses were assigned to four seaplane tenders of

UF-1 141270 shows off its factory-fresh seaplane gray finish. The fuselage band and wing tips are yellow-orange with white edges. (Photo Courtesy of the Grumman Corporation via Hal Andrews)

This perspective of UF-1 141269 shows the upper surfaces detail. The walkway is black with a white outline. (Photo courtesy the Grumman Corporation via Hal Andrews)

SA-16B/ASW 51-050 was one of two evaluated by VP-31 at NAS North Island prior to delivery to Norway. The Albatross is finished in aircraft gray over white in compliance with a USAF Tech order 1-1-4 governing MAP aircraft deliveries. (Photo courtesy Harry Gann via Hal Andrews)

the CURRITUCK Class; they were the USS CURRITUCK (AV-7), USS NORTON SOUND (AV-11), USS PINE ISLAND (AV-12), and the USS SALISBURY SOUND (AV-13). Commissioned during World War II, the PINE ISLAND and SALISBURY SOUND remained on the active rolls until 1971. The CURRITUCK was retired in 1967 while the NORTON SOUND went on to serve as a guided missile ship (AVM). Albatrosses known to have been assigned to seaplane tenders include BuNo 141262 and 142362.

Albatrosses remained operational with the U.S. Navy until 1976 when the last UF-2 (BuNo 141266) was flown from its Guantanamo, Cuba, base to Sherman Field, Pensacola, Florida, where it was turned over to the Naval Aviation Museum.

The U.S. Navy was known for its colorfully marked aircraft and the Albatross was no exception. Over the years the Navy issued a number of directives that governed color and marking standards by aircraft type. Since the Albatross fit into multiple Navy classifications (Patrol, Utility, Seaplane, and SAR), the basic standards were followed. The uniform of the Navy Albatross was divided between a utility scheme and a high visibility search and rescue scheme. When the production Albatross joined the Navy in 1949, the color guidelines in effect for utility aircraft and seaplanes called for an overall glossy sea blue finish. The original directive, dated October 7, 1944, was amended on January 2, 1947, ordering utility amphibians finished in overall aluminum. Within this amendment was a degree of latitude which allowed water-based aircraft to retain their sea blue finish. A similar amendment was issued May 1, 1948, followed in July by the delivery of

Wearing the SAR scheme, UF-1 Bureau Number 131908 awaits its ferry flight to Argentina in Fall 1961. (Photo courtesy U.S. Navy via Hal Andrews)

This view of a UF-1 in a steep bank proves that the Albatross was a well-proportioned airplane from any angle. (Photo courtesy the Grumman Corporation via Hal Andrews)

CHAPTER 8: NAVAL AVIATION • 71

Finished in overall engine gray, UF-1 Bureau Number 137927 is prepared for storage with a rudder block in place and the engine intakes sealed. (Photo courtesy U.S. Navy via Hal Andrews)

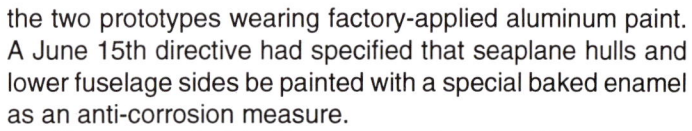

A white and gray HU-16D assigned to the Naval Mission to Columbia in August 1967. Albatrosses assigned mission duty were commonly flown by Marine pilots. (Photo courtesy U.S. Navy via Dave Ostrowski)

the two prototypes wearing factory-applied aluminum paint. A June 15th directive had specified that seaplane hulls and lower fuselage sides be painted with a special baked enamel as an anti-corrosion measure.

Near the end of the Korean war, when the Navy returned to colorful markings, UF-1s were trimmed with orange-yellow floats and struts, wingtips and fuselage wrap-around bands with white borders. On the forward fuselage, a yellow or white rectangular border framed the last three digits of the Bureau Number.

In keeping with a paint scheme modernization program begun during 1955, Navy Albatrosses were ordered painted in overall semigloss seaplane gray; the gloss version of which was adopted in 1961 and known as engine gray. This major changeover was to be in effect by July 1, 1957. In the interim, a provision was introduced in July 1956 for the application of solar heat reflecting gloss insignia white to upper fuselages. A number of Albatrosses remained overall gray since application of the white was at the discretion of unit commanders. During this major revision, the Navy initiated the practice of painting wheel wells white.

As Naval aviation neared the end of the decade, the Bureau of Aeronautics developed a high visibility color configuration that went into effect on February 12, 1959. This scheme had semigloss fluorescent red-orange paint applied to the nose, wingtips, and tail surfaces while movable surfaces and hull bottoms, including the lower fuselage, were exempt. As part of this change, a memorandum directed that landing gear door edges be painted insignia red.

On December 27, 1961, a color scheme specifically for search and rescue was adopted. Included was an orange-yellow rectangle, bordered with black, positioned on the forward fuselage which framed the last three digits of the Bureau Number. This necessitated the repositioning of the national insignia from the bow to the aft fuselage. In addition, a 36 inch orange-yellow band, also edged with black, encircled

HU-16D 141283, based at Kwajalein, wears this colorful Navy scheme adopted in December 1961. Number 283 is seen here at Wake Island in August 1964. (Photo courtesy Dave Menard)

72 • GRUMMAN ALBATROSS: A HISTORY OF THE LEGENDARY SEAPLANE

Left: This UF-1L, Bureau Number 142429, was used by Antarctic Development Squadron Six (VXE-6), which used the tail code "XD." Other markings included a penguin and the fitting words "Shake, rattle and roll" barely visible immediately below the nose anti-glare panel. Both UF-1Ls had a large window added just below the wing trailing edge and both were the only Navy triphibians, which added nearly 700 pounds to the aircraft's basic weight. The dark shape at the aft afterbody is a flush antenna. (Photo courtesy Maj. David H. Brazelton, USAF via Bob Lawson) Right: UF-1 131893 in the sea blue scheme at NAS Kodiak on October 20, 1955. (Photo U.S. Navy via Bob Lawson)

UF-1 137916 flies over Kwajalein Island in March 1956. (Photo courtesy Jack Bradford via Bob Lawson)

HU-16D at NAS Barbers Point during summer 1969. The plane captain's rank and name are painted below the copilot's window. (Photo courtesy Nick Williams)

Number 137927 as it appears fully restored to its original Navy finish and markings when it was assigned to NAS Guantanamo Bay. (Photo courtesy Terry Love)

CHAPTER 8: NAVAL AVIATION • 73

Left: Finished in overall glossy sea blue, the third production UF-1 flies over Norfolk, Virginia in August 1950 during Navy acceptance tests. The letters "FT" below the cockpit denote "Flight Test". An instrument test probe is fitted to the port wing tip. (Photo courtesy U.S. Navy)
Right: A factory-fresh UF-1, Bureau Number 131904, is finished in glossy sea blue. This Albatross thrives today as a corporate-owned aircraft in Florida. (Photo courtesy U.S. Naval Institute)

Left: Seen here on takeoff, Bureau Number 131914 was the first of five UF-1Ts built as a Navy trainer. (Photo courtesy U.S. Naval Institute)
Right: With a minimum of markings, the last operational Albatross, HU-16D 141266, is suspended over a calm sea prior to its final water landing as a USN aircraft. The Albatross was then flown to Sherman Field, Pensacola where it was turned over to the Naval Aviation Museum. It was piloted by CDR Charles Larzelere of Aircraft Ferry/Transport Squadron Thirty one (VFR-31). (Photo courtesy U.S. Navy)

A group of Albatrosses, comprising UF-1s and UF-1Ts, share the ramp at Annapolis with two Grumman JRF **Gooses**. (Photo courtesy the Grumman Corporation)

74 • **GRUMMAN ALBATROSS: A HISTORY OF THE LEGENDARY SEAPLANE**

Left: UF-2 141282 over its home port Naval Air Station Bermuda on August 22, 1964. (Photo courtesy U.S. Navy) Right: Wearing the SAR scheme adopted in late 1961, HU-16D 137919 assigned to NAS Adak, Aleutian Islands, flies over Alaska on January 2, 1965. (Photo courtesy U.S. Navy via Bob Lawson)

the fuselage just forward of the tail. The upper wing section, between the outer edges of the engine nacelles, was painted orange-yellow with a black border. Wing tips received similar treatment and the floats and their struts became orange-yellow as well. After May 20, 1964, the word RESCUE in black was added to the yellow upper wing section. The same ruling directed that the last four Bureau Number digits be painted in white on the hull bottom.

Before ending its Navy career, the Albatross would undergo a final change in livery. A June 29, 1964, directive phased out seaplane gray in favor of light gull gray and further stipulated that all Albatrosses have white upper fuselages.

The major color scheme revision issued in February 1955 introduced guidelines for aircraft operating in arctic regions. For the pair of overall aluminum UF-1Ls assigned to VXE-6, this meant that the wing tips and entire tail section were painted international orange.

Albatrosses that served Naval Attaches and Naval Missions were painted in accordance with a directive issued June 16, 1952. This called for large American flags applied to the tail fin as well as the lower left wing. Whether finished in overall seaplane gray or the white over gray scheme, all displayed large white UNITED STATES NAVY titles on the aft lower fuselage. Those in the split scheme had a black demarcation stripe and engine cowling chevrons added in compliance with a directive dated July 26, 1967.

This UF-2, painted seaplane gray, wears markings in accordance with a Navy directive for aircraft assigned to attache duty. The center cabin window has been replaced by an oval observation blister. (Photo courtesy Jack M. Friell)

CHAPTER 8: NAVAL AVIATION • 75

U.S. NAVY ALBATROSS PRODUCTION/CONVERSION

MODEL	BUREAU NUMBER
XJR2F-1 Prototypes	
UF-1 (HU-16C)	82853 and 82854
	124374 to 124379
	131889 to 131912
	137899 and 137900
	137902
	137905 and 137906
	137908 and 137909
	137912
	137916
	137918
	137921
	137923 to 137930
	137932 and 137933
	141262
	141265
	141267 and 141268
	141271 to 141273
	141276 and 141277
	142360 to 142362

MODEL	BUREAU NUMBER
UF-2 (HU-16D)	131913
	137901
	137903 and 137904
	137907
	137910 and 137911
	137913 to 137915
	137917
	137922
	137931
	141261
	141263 and 141264
	141266
	141269 and 141270
	141274 and 141275
	141278 to 141283
	142358 and 142359
UF-1T (TU-16C)	131914 to 131918
UF-1L (LU-16C)	142428 and 142429

Does not include aircraft assigned Bureau Numbers for MAP

Above: HU-16C 131911 in July 1967. (Photo courtesy Nick Williams)

Above right: HU-16D Bureau Number 141269 in its second scheme at Midway Island. (Photo courtesy Nick Williams)

Right: HU-16C 137924 displays the complete post 1964 Navy SAR scheme. (Photo courtesy Candid Aero Files)

76 • GRUMMAN ALBATROSS: A HISTORY OF THE LEGENDARY SEAPLANE

CHAPTER 9

WORLDWIDE SERVICE

Besides Albatrosses manufactured by Grumman specifically for the governments of Japan, Canada, West Germany, and Indonesia, many were supplied to foreign military forces under the U.S. Military Assistance Program (MAP), or the Mutual Defense Assistance Program (MDAP). A number of aircraft were also rebuilt by Grumman to specifications submitted by foreign governments, usually in the form of ASW configurations. With the resurgence of popularity in the Albatross, some restoration firms are turning to South American countries to purchase Albatrosses, most of which are reported to be in remarkably good condition.

In accordance with a 1961 MAP contract with Argentina, five former USAF SA-16As were upgraded by Grumman to SA-16B standards equivalent to the UF-2G (HU-16E) of the U.S. Coast Guard. Two were placed in service with the Navy and supplemented by another pair transferred directly from USAF stocks, bringing Argentina's total to seven aircraft. Four were first operated by the Escuadrilla Aeronaval de Busqueda y Salvamento (SAR Squadron) at NAS Punta del Indio. Naval Albatrosses, designated 4-BS-1 through 4-BS-4, were later assigned to the 2nd Propositos Generales Squadron at NAS Commandante Espora and the Naval SAR School where they served until 1978. The remaining three aircraft, numbered BS-01 through BS-03, went to the Fuerga Aerea (Air Force) for duty with the SAR squadron at Air Station El Palomar, Buenos Aires. During December 1972, one of the trio began flying for Lineas Aereas del Estado, a military airline comprising civil and military crews for regular service from Comodoro Rivadavia to Port Stanley in the Falklands. Two of the Air Force aircraft were temporarily equipped with triphibian gear during the late 1960s by the Antarctic Air Task Force (FATA) for flights between Rio Gallegos and two Antarctic stations; Vice Comodoro Marimbio and Benjamin Matienzo. At least one Albatross remained in service for utility and search and rescue until the early 1980s.

In 1959, the U.S. Air Force transferred 14 SA-16As to the Brazilian Air Force (FAB) where they were assigned serials 6530 through 6543. The amphibians were initially designated A-16, then U-16, M-16 in 1976 and S-16 in 1977. At that time ten remained in SAR service and number 6541 was written off in December. The search and rescue group at Florianopolis operated the S-16s until late August 1980. Number 6530 later went to the air museum and numbers 6528, 6529, 6532, 6533, and 6538 were scrapped at Sao Paulo.

Ten Albatrosses were built for the Royal Canadian Air Force (RCAF) with the first example delivered on July 20, 1960. Based on the SA-16B, these aircraft were designated CSR-110s in Canadian service and assigned serials 60-9301 through 60-9310. The RCAF order was completed by November 1960 and all were delivered with triphibian gear.

Left: Although given Navy Bureau Numbers for the transfer, two HU-16Bs were delivered under a MDAP contract to the Royal Thai Navy in 1968. At least one of these aircraft, pictured here, is known to have been delivered from the 304th ARRS at Portland, Oregon. (Photo courtesy David Wendt) Right: An out of service and badly weathered HU-16A, 51-037, in 1987. (Photo courtesy Terry Love)

CHAPTER 9: WORLDWIDE SERVICE • 77

Left: The West German Navy operated a total of eight Albatrosses which underwent three designation changes throughout their service. This is one of a group of five aircraft built under MAP contract which were based on the UF-2. Number 60-08 is at Kiel in 1971. (Photo courtesy MAP via John Lameck) Right: A total of twelve Albatrosses flew under the flag of the Philippine government. This HU-16B appears at Long Beach in May 1977. (Photo courtesy Nick Williams)

Unique to the Canadian Albatrosses was the installation of two Canadian-built Wright R-1820-82 engines. The 1,525 hp powerplant was easily identified by air intakes atop the engine cowlings. Another distinguishing feature of the CSR-110s was modifications to the retractable undercarriage to facilitate beaching. The CSR-110s became operational with the 442nd Transport and Rescue Squadron at Comox, British Columbia. They were also flown from Greenwood, Nova Scotia; Trenton, Ontario; and Winnipeg, Manitoba, until they were phased out in 1970. Eight of the ten aircraft went on to serve the air arms of Chile and Mexico, while the remaining two passed into commercial ownership.

Three SA-16As (S/N 49-097, 49-099, and 49-100) were transferred from the U.S. Air Force to the Chilean Air Force in 1958. These were later converted to ASW SHU-16Bs and augmented by three additional SHU-16Bs (S/N 51-014, 51-024, and 51-7191) supplied under MAP/MDAP in 1963. Number 51-7191 was originally intended for Columbia and painted

The second CSR-110 built for the RCAF leaves the Grumman facility for Trenton, Ontario. (Photo courtesy the Grumman Corporation via Hal Andrews)

in Columbian Air Force markings before delivery to Chile. The six Albatrosses, assigned serial numbers 566 through 571, were operated by the 2nd Group at Santiago-Los Cerillos until 1979. In 1971, Chile's Naval Aviation Service acquired

This is the fifth of six UF-2s delivered under MAP to Japan in 1961. Assigned USN Bureau Number 148328, this Albatross is painted sea blue. The air intakes atop the engine cowlings identify the installation of the more powerful R-1820-82 engines. This Model G-262 flies over Long Island, New York, prior to delivery. (Photo courtesy the Grumman Corporation via Hal Andrews)

78 • GRUMMAN ALBATROSS: A HISTORY OF THE LEGENDARY SEAPLANE

A West German Bundesmarine Albatross taxis up a seaplane ramp. The yellow-orange continues around the hull and the pilot's escape hatches are the aircraft color. (Photo courtesy MFG-5)

A Bundesmarine Albatross water taxis while a crewman in the starboard hatch prepares to mount a JATO bottle for takeoff. (Photo courtesy MFG-5)

four ex-Canadian CSR-110s (9301, 9302, 9303, and 9310) which served in the SAR role until 1979. Numbers 9303 and 9310 were purchased by Grumman for modification and re-sold to Malaysia, while another became a Chilean museum display, and the fourth was written off in November 1973. The first two aircraft procured by Chile ended up on civil registry, one going to the Confederate Air Force.

Based on the UF-2, five Albatrosses were manufactured for West Germany with the first unit delivered on 8 January 1956 and the last on April 20th. Under MAP contract, these aircraft were given U.S. Navy bureau numbers 146426 through 146430 but numbered SC101 through SC105 in Kriegsmarine service.

Marinefliegergeschwader 5 (MFG 5) was formed at Kiel-Holtenau in 1958 for search and rescue and radio relay duties with the five Albatrosses. These were later supplemented by three former USAF SA-16As (S/N 49-088, 49-095, and 49-096). Later, in Bundesmarine service, the eight aircraft became RE501 to RE508 and finally designated 6001 to 6008 in January 1968. In 1972 the oldest of the group, 49-088, was scrapped while the remaining seven were sold on the U.S. civil market.

In 1970 Greece received twelve ASW SHU-16Bs from Norway. Originally procured under MAP, these aircraft were assigned serial numbers 51-044, 51-068, 51-070, 51-5289, 51-7177, 51-7183, 51-7196, 51-7201 to 51-7204 and 51-7207. A former Spanish SHU-16B (S/N 51-5283) was obtained later.

A West German Bundesmarine Albatross with gear down approaches a seaplane ramp. (Photo courtesy MFG-5)

Delivered as Bureau number 146430, Bundesmarine Albatross number 6005 clears the water following a JATO takeoff. (Photo courtesy MFG-5)

Greece also received an HU-16C (BuNo 137909) as a source of spare parts. An HU-16D was also purported to have been used for parts, however it seems that BuNo. 137915 averted this demise since it thrives on the current civil register. As of late 1993, ten Albatrosses were operational with 353 Squadron flying utility missions with the Greek Air Force. One aircraft was turned over to a Greek museum.

Left: Using a single JATO unit on each side, a West German Albatross becomes airborne. (Photo courtesy MFG-5) Right: The first Albatross delivered under MAP to West Germany. Number 101 was assigned Bureau Number 146426. (Photo courtesy MFG-5)

CHAPTER 9: WORLDWIDE SERVICE • 79

Left: The first five Albatrosses built for the West German Navy were based on the UF-2 and assigned USN Bureau Numbers. Germany's amphibians were painted overall dull aluminum. Number 103, seen here taxiing on the water, has yellow-orange trim. The letters SAR are superimposed on both wing bands. The German national fin flash is black/red/gold. (Photo courtesy MFG-5) Right: Number 102 of the West German Navy in early markings which included a fin flash in the national colors of black/red/gold, an orange fuselage band aft of the wings, and the Kriegsmarine emblem below the cockpit. The overall finish was dull aluminum. (Photo courtesy the Grumman Corporation via Hal Andrews)

The Icelandic Coast Guard acquired two HU-16Cs (BuNo 141276 and 142361) on a lend-lease program in 1969. The type was never adopted for service use and both Albatrosses were fully restored and are currently flown by private owners.

The Air Force of the Indonesian National Armed Forces was the first foreign air arm to purchase Albatrosses direct from Grumman. Similar to late production UF-1s, eight examples were delivered during 1957 and 1958 and serialed PB/517 to PB/524. Later supplements included two former U.S. Navy UF-2s (BuNo 137907 and 142359), Four German aircraft (6005 to 6008) and a former Japanese variant, (BuNo 148329) numbered as 9056 in Japanese service, were also added to the Indonesian fleet. Grumman records indicate that two aircraft were converted to ASW SHU-16Bs during the early 1960s which were probably among the Albatrosses later transferred to the Naval Aviation Branch.

In 1958 the Italian Air Force obtained six former USAF SA-16As with serials 50-174, 50-175, 50-177, 50-179, 50-180, and 50-182 to equip the 140th Squadron of the 84th Group at Vigna di Valle. The role of the 84th Group expanded when it was absorbed into the 15th Stormo in 1965 and another six HU-16As were added with serials 51-035, 51-037, 51-7157, 51-7175, 51-7252, and 51-7253. In Italian service, Albatross serial numbers were prefixed by "MM" which represented Matricola Militare. The fate of the Italian amphibians was sealed during the late 1970s when the company responsible for overhauling the type folded and the Italian Air Force decided to retire all radial-engine aircraft.

A total of six UF-2s, equipped with the more powerful R-1820-82 engines, were built for Japan under MAP and delivered to the Japanese Maritime Self Defense Force (JMSDF). Assigned BuNo.s 148324 through 148329, the first aircraft was delivered on 17 February 1961 with the final delivery completed on May 5th. In Japanese service the amphibians were numbered 9051 through 9056 and assigned to the Omura Kokutai until 1976 when five were sold on the civil market and converted to G-111s.

The Royal Malaysian Air Force became the third air arm to operate two Albatrosses that started life as Canadian CSR-110s numbered 9303 and 9310. After also serving in Chile,

With a color scheme matching Coast Guard Albatrosses, this "B" Model carries a civil registry with its Philippine Air Force markings. (Photo courtesy Terry Love)

Originally designated AN.1A-10, this SHU-16B, was obtained from Norway in 1969. Note the unit emblem on the nose and the prop oil cuff colored to match the national insignia. (Photo courtesy Candid Aero Files)

80 • GRUMMAN ALBATROSS: A HISTORY OF THE LEGENDARY SEAPLANE

Left: This JMSDF UF-2 was assigned USN B/N 148325 and is painted in the later scheme of white and engine gray with orange trim. Note the landing gear in the final retraction cycle following takeoff. (Photo courtesy Candid Aero Files) Right: HU-16B of the Indonesian Air Force at NAS Alameda in mid 1977. (Photo courtesy Terry Love)

the pair was refurbished by Grumman before delivery to Malaysia in 1986. Number 9303 was customized with an executive interior for the Prime Minister's use. They received RMAF serials M35-01 and M35-02 and are believed to be in current operation.

The Mexican Navy took delivery of four Canadian CSR-110s (9305, 9306, 9307, and 9309) to replace six Beechcraft C-45s and equip its 3rd Naval Air Squadron with detachments at Ensenada and La Paz. After refurbishment by Grumman, the four were delivered in 1974 and numbered MP-101 through MP-104. In April 1976, they were supplemented by nine former U.S. Navy HU-16Cs: bureau numbers 137913, 137914, 137917, 137920, 141270, 141273, 141274, 141280, and 141283. They were in storage before being sold to the Mexican Navy and delivered in early 1979. Another ex-RCAF CSR-110 (9308) was also acquired, but not before it was converted to G-111 standards. The ten aircraft were numbered in 300 and 400 series numbers prefixed by "MP." At least two Albatrosses have found their way to the American commercial market while some may still be operational with the Mexican government.

As the largest operator of ASW Albatrosses, the Royal Norwegian Air Force obtained 18 SHU-16Bs to equip its 330th and 333rd anti-submarine patrol squadrons. Deliveries began during May 1961 and were completed by March 1964. Included as a MAP procurement were USAF aircraft with

This HU-16A, serial number 50-174, was the first Albatross delivered to the Italian Air Force in 1958. The emblem of the 15th Stormo is carried on the tail fin and the forward keel is painted yellow. Note the inability of fluorescent red-orange to withstand weather. (Photo courtesy Terry Love)

CHAPTER 9: WORLDWIDE SERVICE • 81

Left: HU-16B number 0533/4-BS-1 of the Argentine Navy's SAR Squadron at Commandante Espora NAS in 1969. The hull is painted black up to the water line and the rudder is patterned after the Argentine flag. The squadron badge is worn on the nose and a triangular flag below the cockpit represents the Squadron Commander's aircraft. (Photo courtesy Jorge Felix Nunez Padin) **Right:** One of three HU-16Bs used by the Argentine Air Force seen at MacDill AFB, Florida, in January 1963. (Photo courtesy Bob Mikesh via Nick Williams)

serials 51-040, 51-044, 51-048, 51-050, 51-060, 51-068, 51-070, 51-474, 51-5281, 51-5283, 51-5288, 51-5300, 51-7177, 51-7183, 51-7190, and 51-7201 through 51-7207. When the aircraft were replaced in 1969, ten went to Greece, six to Spain, and two to Peru.

Four former USAF SA-16As were supplied under MAP to the Pakistan Air Force in 1958 and flown by No. 12 Squadron. Later they were assigned to No. 4 Squadron until phased out during the early 1970s.

As a MAP procurement, Peru received three SHU-16Bs (51-038, 51-041, and 51-7174) for its Air Force in 1963. They were supplemented by two ASW Albatrosses (51-048 and 51-050) obtained from Norway in 1970. They were numbered 517 through 521 (number 518 was later written off) and assigned to the 31st Group at Lima-Callao for maritime reconnaissance and search and rescue until retirement during the early 1980s.

The Philippines Air Force obtained former U.S. military Albatrosses through MAP from a number of sources beginning during the mid 1950s with four USAF SA-16As serialed 48-589, 48-599, 48-605, and 48-607. During the 1970s, four USAF HU-16Bs were added having serials 51-019, 51-473, 51-7151, and 51-7184. Added to that list were two former U.S. Coast Guard HU-16Es (1267 and 7248) and finally, two HU-16Cs from the U.S. Navy (BuNo 137906 and 137922).

On December 4, 1962, two U.S. Coast Guard pilots stationed at NAS Sangley Point, Philippines, were detailed to test fly BuNo 137907 at the Manila airport. Philippine Airlines overhauled the aircraft and converted it into an unusual configuration that included gun ports in the cabin. Much of the usual equipment, such as de-icing gear, electronics, and radar had been removed resulting in a very light aircraft. The amphibious oddity did not meet the Coast Guard preflight criteria and after several failed attempts, the wary pilots finally got it airborne. While the intended purpose of this aircraft is unknown, remaining Philippine Albatrosses performed search and rescue duties. Some were still in service with the 27th SAR and Reconnaissance Squadron at Sangley Point near the end of the 1980s.

Under MDAP, the Portuguese Air Force took delivery of three SA-16As with serials 51-5277, 51-15270, and 51-15271. These were assigned to the 4th Squadron at Lajes, Azores, until phased out in 1962.

Spain bought the three Albatrosses from Portugal which brought the Spanish inventory up to fifteen amphibians. Five SA-16As had been supplied under MDAP to the Spanish Air

SA-16A number 51-15271, seen here preparing for its test flight at the Grumman plant, was one of three delivered to the Portuguese Air Force. (Photo courtesy the Grumman Corporation via Dave Ostrowski)

Following service as a Canadian CSR-110, this Albatross was one of four renovated by Grumman for the Mexican Navy. (Photo courtesy MAP)

82 • GRUMMAN ALBATROSS: A HISTORY OF THE LEGENDARY SEAPLANE

The RCAF scheme seemed to accentuate the inherent beauty of the Albatross. Number 9301 is seen here at RCAF station, Trenton, on July 1, 1961. (Photo courtesy Larry Milberry)

Finished in aircraft gray over white, this SHU-16B was one of 18 examples that served the Royal Norweigan Air Force during the 1960s. (Photo courtesy Jack M. Friell)

Force in 1954 and supplemented by seven HU-16As and HU-16Bs during the early 1960s. The earlier "A" models were prefixed by AD.1 and the HU-16Bs AD.1B.

They were assigned to two SAR units; the 55 Escadrille which was redesignated 55 Squadron in 1963 and 801 Squadron in 1967 at Son San Juan, Palma de Mallorca, and the 56 Escadrille which became 56 Squadron and later 802 Squadron at Gando in the Canary Islands. In 1963, seven SHU-16Bs, serialed 51-069, 51-7147, 51-7148, 51-7165, 51-7167, 51-7170, and 51-7172, were obtained and renumbered AN.1A-1 to AN.1A-7. An additional six ASW aircraft were obtained from Norway in 1969 which became AN.1A-8 to AN.1A-13. All were based at Jeres de la Frontera with 611 Squadron which was redesignated 206 Squadron in 1967 and 211 Squadron in 1973. By 1978 the Spanish ASW Albatrosses were phased out and the remainder replaced by CASA 212s.

The Republic of China Air Force, formerly the Chinese Nationalist Air Force, received its first Albatrosses from the USAF in 1958. A total of fourteen SA-16As were transferred to China along with at least three HU-16Bs. Most saw service with the Air Rescue squadron at Chiayi where they performed SAR duties alongside helicopters until replacement in 1987 and 1988. One SA-16A, numbered 11024, is displayed at the Chung Cheng Aviation Museum in Taiwan. During a rescue mission off the Island of Matsu in 1966, a SA-16A was shot down by a Mig of the People's Republic of China.

A pair of HU-16Bs, formerly flown by the USAF, was transferred under MDAP to the Royal Thai Navy in 1968. For the transfer, they were assigned U.S. Navy Bureau Numbers 151264 and 151265. The latter was serialed 51-7235 in USAF service and both were used for SAR duties until retirement in 1981.

Though difficult to substantiate, six former USAF HU-16As are purported to have been in use with the Venezuelan Navy for nearly two decades before being phased out in 1982.

Number 568 was one of three subhunter variants acquired by the Chilean Air Force in 1963. The emblem below the cockpit reads, "Grupo No. 2 ASW." (Photo courtesy Jorge Padin)

CHAPTER 9: WORLDWIDE SERVICE • 83

CHAPTER 10
"TT" AIRLINES

The Albatross commenced an early civil career with the ninth, eleventh, and twentieth aircraft built as SA-16As with serial numbers 48-596, 48-597, and 48-603. In Fall 1955 they were transferred from Air Force service to the Department of the Interior and given civil registration numbers N9942F, N9943F, and N9944F respectively. The trio, plus two Douglas DC-4s, provided air service to the U.S.-administered Trust Territory comprising the Mariana, Caroline, and Marshall Islands in the Western Pacific, also known as Micronesia. The exotic South Seas routes connected 96 inhabited islands and atolls amidst three million square miles of ocean dotted with over 2,000 islands, whose combined land mass is less than the state of Rhode Island.

The alluring visions of pilots who flew these routes quickly dissipated when they realized that the beauty of the islands masked primitive conditions and oppressive heat. U.S.-sponsored air service actually began in 1949 under contract with Transocean Airlines with converted PBY-5 "Catalinas." After receipt of the three Albatrosses, which were modified to carry fifteen passengers, four crewmembers, and 600 pounds of mail, Transocean flew the island routes for five years before going bankrupt.

In July 1960, the Trust Territory Air Service, which the pilots nicknamed "TT" Airlines, and the five aircraft, were taken over by Pan American World Airways. This arrangement prospered with the aircraft flown by Pan Am pilots, serviced by Pan Am mechanics, and passengers booked through local government people. Pan Am's maintenance people were hard-pressed to keep the SA-16As fully operable, especially since parts were difficult to obtain. Using ingenuity, Pan Am's main-

One of three SA-16As transferred from the U.S. Air Force to the Department of the Interior for use by Transocean Airlines. Number N9943F wears the Trust Territory seal below the cockpit alongside Transocean's wing logo. The white over dark blue color scheme sported yellow upper engine nacelles, tail number, and logo wings. (Photo courtesy W.J. Balogh via Dave Menard)

Both HU-16Bs operated by Pan Am to carry U.S. government personnel to and from the satellite tracking station at Mahe were marked with the Pan Am logo, American flag, and USAF serial. (Photo courtesy D. Colbert via Dave Menard)

tenance supervisor on Guam overcame the problem by keeping track of Albatross wrecks. He was advised by friends and the Coast Guard of wrecks off the Oregon and California shores, plus a Navy UF-1 lost during the 1960s on Chi Chi Jima. Two sank in Apra Harbor, Guam, which were pulled ashore and cannibalized for parts with the hulks used by the Navy for firefighting practice. Highly corrosive salt water was a constant problem alleviated only by fresh water washdowns. High humidity affected ignition systems, which were later rubber-sealed and magnetos which were modified with air valves and purged with dry nitrogen.

During Pan Am's eight year tenure as TT Airlines, the Albatrosses, nicknamed "Clipper Ducks" by the pilots, flew 250 search missions for missing boats, made 150 landings in reef-laced lagoons to deliver doctors or take out severely ill islanders, and performed ten medical evacuation missions following major typhoons.

Airport facilities were eventually upgraded for jet operation and a new airline, Air Micronesia, secured the TT contract, making its inaugural flight on May 16, 1968. Some islands were still unable to accommodate jet aircraft, which prompted Air Micronesia to lease two Albatrosses for the Truk-Pohnpei run. By 1970, a new runway was opened at Pohnpei while other improvements were made, which spelled the end of the TT Albatrosses — both were retired in 1972.

During the early 1960s, Pan Am again operated an Albatross (S/N 51-7163) transferred from the Air Force for a joint contract with the U.S. government and the Philco Radio Corporation. Under the call sign "PA 163", the HU-16B flew a weekly service between the satellite tracking station at Mahe in the Seychelles and Mombassa's Port Reitz Airport in Kenya. The first Albatross chosen for the 13 hour flights experienced engine trouble and crashed into the Red Sea on June 28, 1963. The crew of four clung to the downed aircraft for 15 hours before rescue by a British destroyer. In November 1971, PA 163 arrived in the U.S. where it was eventually sold on the civil market and registered N70725. It was replaced by another HU-16B (S/N 51-7219) which arrived from the U.S. in late October 1971. The service continued until late 1972 when a new airport was opened at Mahe and the Albatross became a part of Pan Am's glorious past.

The Trust Territory seal was applied beneath the cockpit of Albatrosses flown by Transocean Airlines and Pan American World Airways in Micronesia.

CHAPTER 11

ALBATROSS EXPLORER

One would think that after many years of service, an airplane relegated to a museum was essentially used up and subsequently "put out to pasture." That appeared to be the case when a Navy UF-2 was retired in 1976 and put on display at the Naval Aviation Museum at NAS Pensacola, Florida. However, the Albatross was recalled to service in 1979 when the museum agreed to loan it to the Smithsonian's National Museum of Natural History, which required a large amphibian for ocean reef research. Bureau Number 141266 was flown to Grumman's rework facility at Stuart Field, Florida, in November for refurbishment and set out on its reconnaissance of Atlantic and Caribbean exploration on January 16, 1980.

Through a business proposition that benefitted all agencies concerned, the Coast Guard donated eleven HU-16Es to the Smithsonian, which traded seven back to Grumman in exchange for the renovation of two. The second Albatross was a UF-2 assigned Bureau Number 146426, which was originally procured for the German Navy and later sold on the U.S. civil market. After modification, the aircraft were registered as N693S and N695S, respectively. Both were used for a four and a half year exploration project sponsored by the Coral Reef Museum of Natural History in Washington, D.C.

With plans to expand its ocean reef exploration worldwide, the Smithsonian's Marine Systems Laboratory had a third Albatross renovated. This aircraft was former U.S. Coast Guard UF-2G number 7249, which was overhauled at Grumman's St. Augustine facility. It is currently operational with the Tropical Research Institution and, as such, is equipped with sophisticated communications gear, a depth finder, and special windows for aerial photography. It, along with two predecessors, have flown reconnaissance, rotated

The first Albatross acquired by the Smithsonian is seen here immediately following renovation. Finished in a high visibility scheme, the UF-2 carries a pair of 150 gallon drop tanks. (Photo courtesy Terry Love)

Photographed in April 1982, N695S shows off her distinctive Smithsonian wing markings at Andrews AFB, Maryland. (Photo courtesy Eugene Zorn via Dave Ostrowski)

personnel, and transported equipment and countless fish, animal, and plant species to the museum's aquarium in Washington.

Selected as the Chief pilot of the Smithsonian's Albatross fleet was Louis Petersen (USN Retired), who flew a Navy Grumman J2F "Duck" from which the historic Japanese surrender in Tokyo Bay was photographed.

Following warm water Pacific reef exploration, the Albatross explorers moved on to explore cold water reefs off the coasts of Maine and Labrador. Also planned were expeditions to British Columbia, Alaska, the Red Sea, India, and Tierra del Fuego, where reef lagoons and fjords provide natural havens for Grumman's seabird.

Still in Coast Guard livery during May 1979, UF-2G number 7249 wears Smithsonian markings shortly after it was acquired to expand the Institution's ocean reef exploration. It was eventually overhauled and modified for use with the Tropical Research Institute. (Photo courtesy Steve Miller)

CHAPTER 11: ALBATROSS EXPLORER • 87

CHAPTER 12
CHALK'S LEGACY AND THE G-111

After thirty years of outstanding military service, the Albatross was redesigned by Resorts International as a commuter airliner. The firm originally leased the revamped amphibians to its Chalk's International Airlines Division for routes linking Miami, Ft. Lauderdale, and West Palm Beach, Florida, and the Bahamian Islands of Bimini and Cat Cay as well as its hotels and casinos at Paradise Island, Nassau.

Billed as the oldest airline, Chalk's had its humble beginnings in 1917. The success of founder A.B. Chalk's airline was fueled by the lucrative smuggling trade during prohibition in 1919, when rumrunners flew to their Bahama haunts. When he retired in 1966, A.B. Chalk sold the unique airline to a friend, leaving behind a legacy rich with tales of adventure, enhanced by Miami's popularity as a vacation mecca. The ability to turn any patch of water into a runway provided an escape for a clientele that included notables Ernest Hemingway, Howard Hughes, Lana Turner, Ava Gardner, Errol Flynn, Judy Garland, and Al Capone, not to mention Havana-bound hijackers.

Chalk's was purchased in 1974 by Resorts International's James Crosby, who updated the fleet of Mallards and Albatrosses with turbine engines — the first turbine-powered Mallard was flown in November 1979. Resorts International then contracted with Grumman to take on the job of transforming military search and rescue Albatrosses into modern airliners.

Designated Model G-111s, the first six were modified at Grumman's facility at Stuart, Florida, while the remaining seven underwent rework at the St. Augustine site. The Albatrosses, culled from a variety of sources, included five UF-2s built for Japan (Model G-262), two Canadian CSR-110s

The first G-111 built makes a water takeoff. Visible on the underside of the engine nacelle is a dark anti-glare panel to ward off reflections from the gloss white finish. Below that is one of two passenger doors added to the forward cabin. (Photo courtesy Grumman Corp.)

88 • GRUMMAN ALBATROSS: A HISTORY OF THE LEGENDARY SEAPLANE

Finished in Chalk's later color scheme, this G-111, registered N120FB, was originally a USAF HU-16B serialed 51-7243. It also served as a USCG UF-2G. (Photo courtesy Lonny McClung of Chalk's International Airlines)

A Chalk's G-111 in its current livery seen at Ft. Lauderdale, Florida, in 1992. (Photo courtesy Udo Schaefer)

(Model G-231), four USAF HU-16Bs, and two USN HU-16Ds. Re-manufacture of the aircraft to G-111 standards took about fourteen months and involved five major phases.

It was said that the Albatross was the amphibious equivalent of the famed DC-3, and like the DC-3, the Albatross was built with materials three to four times stronger than needed, ensuring a long life. A limiting factor in the Albatross' service life however, was its susceptibility to salt water corrosion. Therefore, a crucial segment of the modification program entailed thorough inspection for corrosion and the installation of corrosion-proof titanium spar caps over the critically important box beams to which the wings attached. In addition, all control surfaces were covered with a noncorroding substance.

JATO brackets and wiring were removed and drop tank plumbing and wiring were capped. The rear passenger door was made to open outward and two additional passenger doors were added to the forward cabin.

The Wright radial powerplants were completely overhauled and "zero-timed" (made factory fresh). Modifications in the flight deck brought major improvements in avionics, electrical systems, and pilot comfort. Removal of nearly a ton of vintage electronic components made room for state of the art avionics. Pilots who recall how noisy the Albatross was would be amazed at the low noise levels achieved with carpeting and soundproofing. The 28-passenger interior can be arranged as a combination commuter-utility configuration. With a payload capability of nearly 8,000 pounds, the G-111 easily qualifies as a freight hauler.

The first experimental variant completed the complex renovation with its first flight on February 13, 1979. This prototype G-111 lifted from the West Palm Beach runway in May 1981 on its way to the Paris Air Show for its civil debut. Piloting the plane was Fred Rowley who flew the first test flights in 1947, and Frank Steven who commanded the RCAF Albatross squadron for ten years. Upon returning from Paris, the

Number N117FB served the American and West German navies before it became part of Chalk's inventory. (Photo courtesy Candid Aero-Files)

CHAPTER 12: CHALK'S LEGACY AND THE G-111 • 89

Albatross was awarded an airline transport category certificate by the Federal Aviation Administration (FAA), the first for an amphibian in more than 30 years (those operated by the Department of the Interior in the Pacific were not fully certified).

For landplane operations, the G-111 had a maximum takeoff weight of 30,000 pounds and 31,400 for water operations. Such restrictions seem insignificant when compared to the prototype's weight of 36,000 pounds for the Paris trip, not to mention RCAF Albatrosses often flown at weights up to 39,000 pounds.

Wearing registration number N112FB, the first G-111 was joined by another, both of which began service in early 1982 with Chalk's International Airline.

After Crosby's death in 1986, ownership of the business passed to his sisters, then to Donald Trump, and finally Merv Griffin, all of whom treated the airline as nothing more than a passing business investment. The airline took a downward turn as Trump down-sized the operation to four Mallards and "mothballed" all the Albatrosses at Bob's Air Park in Arizona. They remained in storage until a change in ownership in 1991 breathed new life into the company. The privately owned United Capitol Corporation purchased the airline and returned three aircraft to commuter service in 1993 following a comprehensive inspection and repair program. With a renewed appreciation and interest in seaplane aviation, Chalk's not only revamped its aircraft, but bought new ones to expand its service. Current expansion plans include positioning by Chalk's to resume flights to Cuba's oceanside resorts as economic relations between the U.S. and Cuba improve. One thing that never changed was the Albatross' versatility, affirmed when Chalk's used them to evacuate the islands and conduct searches for survivors during 1992's Hurricane Andrew.

Resorts International's venture with the G-111 boosted Grumman's anticipation of a strong market for the "new" Albatross. By 1983, the company had re-purchased 36 aircraft from government sales, ten from a broker, four from Chile, and seven from the Smithsonian Institution. Resorts International initially purchased four of those and ordered another seven for delivery in 1983. Many of the government surplus aircraft were destroyed by a violent storm at the MASDC and necessarily scrapped. The remainder were stored at Bob's Air Park in Arizona awaiting a market that never materialized. Not to be discouraged, Grumman St. Augustine, which opened in 1980, continues to stock over 15,000 items of Albatross components in anticipation of still active sales. Other markets Grumman hopes to arouse interest in center around offshore oil rigs and island routes in Hawaii and the Caribbean.

Of 13 G-111s produced, 12 are owned by Chalk's along with three HU-16Bs; the 13th G-111 is privately owned. Grumman re-purchased the first production G-111, registered N114FB, and resold it to Pelita Air Services of Indonesia, who in turn leased it to Conoco for offshore oil platform work under the registration PK-PAM. This Albatross was originally the last of six Model G-262s built for the Japanese Navy and assigned Bureau Number 148329. Built in 1961, it was the last Albatross to roll off the production line. After testing HU-16s borrowed from the Indonesian Navy, Pelita acquired a number of them from the Indonesian government and had them converted to 20-passenger airliners. PK-PAM operated out of Singapore and Indonesia until purchased by the Paragon Ranch in Colorado where it was registered as N26PR and put up for sale in April 1993.

The success of the G-111 has also contributed to the prospective development of a "water bomber." The firefighting G-111 would rapidly fill twin tanks in the hull with five tons of water, drawn through retractable scoops as the aircraft skims the surface of a nearby body of water. Over the fire, the water could be released separately from each tank or simultaneously.

The status of Chalk's G-111 is again undergoing change as the firm prepares to convert its Albatrosses to turboprop variants called G-111Ts.

CHALK'S INTERNATIONAL AIRLINES/FLYING BOAT INC. 1993 ALBATROSS AIRCRAFT INVENTORY

CIVIL REG. No.	MILITARY No.	MILITARY DES.	MILITARY SERVICE
N112FB G-111	148328	UF-2	Germany
N113FB	51-7244	HU-16B/E	USAF/USCG
N115FB	148327	UF-2	Japan
NI16FB G-111	148325	UF-2	Japan
N118FB G-111	60-9304	CSR-110	RCAF
N119FB G-111	60-9308	CSR-110	RCAF
N120FB G-111T	7243	UF-2G	USCG
N121FB G-111	7249	UF-2G	USCG
N122FB	51-7168	HU-16B	USAF
N124FB G-111	137901	UF-2	USN
N125FB	141282	UF-2	USN
N695S	146426	UF-2	Germany
N114FB	1311	HU-16E	USCG
N117FB	148326	UF-2	Germany

CHAPTER 13

WARBIRDS AND PLEASURE CRAFT

Commercial use of the Albatross seems to have had its beginnings in 1970 when Grumman conducted a feasibility study to alleviate the growing commuter problem in New York City. The company proposed a plan using turbo-powered Mallards and Albatrosses to handle up to 5,000 passengers daily, operating near New York's financial district. Little came of the study and Grumman concentrated its energies elsewhere.

Towards the end of the decade, four commercial Albatrosses supplemented the amphibious fleet of Antilles Air Boats in the Caribbean before the company was acquired by Resorts International. Also plying Caribbean waters were a number of former military Albatrosses involved in drug smuggling. Albatrosses were among the many aircraft scattered throughout the islands, abandoned after serving their sinister purpose in Caribbean drug operations. The aircraft, some intentionally stripped of their log books and identification plates, were confiscated and sold at auctions. These, along with other surplus Albatrosses parked at MASDC in the Arizona desert, found their way into the hands of private owners and collectors.

During the 1980s and 1990s, the Albatross enjoyed a resurgence in popularity on the civil market due to the inherent integrity of the design, its versatility, plus its range and payload. By the end of 1993, 84 Albatrosses appeared on the U.S. civil register alone. Flying Boat Inc. and South Florida Aviation Investments Inc. at Opa Locka share ownership of more than one third of that number.

During the 1980s, the amphibians were also undergoing rejuvenation at Canada's Victoria, British Columbia, airport for use as charter aircraft on the waterways of Canada and Alaska.

In the U.S., Grumman St. Augustine continues support of more than thirty Albatrosses from Chalk's to those of the Greek Air Force. Some of Grumman's customers have brought back Albatrosses from South American countries, Mexico, and Puerto Rico. Three were sold recently by the Drug Enforcement Agency which confiscated the aircraft in Texas, Florida, and Grand Turk Island.

Displaying a stark contrast in color schemes, two restored Navy Albatrosses fly in formation. The 'warbird', number 131911 was a UF-1 while the white bird was a UF-2 with service in the German and Indonesian military. (Photo courtesy Steve Penning)

Left: Registered as N46RG, this Albatross has "KATANGA" painted below the rear cabin windows. The name was a carryover from the aircraft's USAF service as HU-16B number 51-7169. (Photo courtesy Steve Penning) Right: Former UF-1 137928 undergoes preparation for its first flight in 25 years at MASDC in June 1993. (Photo courtesy Steve Penning)

CHAPTER 13: WARBIRDS AND PLEASURE CRAFT • 91

Left: Aerocrafters workers mount a rebuilt engine to Richard Sugden's UF-1 at MASDC in November 1993. Before the Albatross was flown to Aerocrafters for overhaul, registration number N7026J was spray painted on the aft fuselage as a FAA requirement. (Photo courtesy of Steve Penning) Right: Two restored Navy Albatrosses, 131911 and 142361, make an impressive warbird display at Santa Rosa airport in November 1992. (Photo courtesy Steve Penning)

A handful of agencies and individuals lead the field in Albatross restoration. After spotting UF-1 number 137927 advertised for sale in a surplus catalog in 1980, Dennis Buehn managed to purchase the aircraft in 1985. He restored the aircraft in the Navy scheme worn during the days when he earned his aircrew wings in the same aircraft. He went on to obtain six additional Albatrosses.

Kirk Williams, of the Paragon Ranch in Colorado, acquired the last Albatross produced, which he put up for sale. He has also purchased HU-16B number 51-7169 as well as three SA-16As from the Brazilian Air Force.

Aerocrafters specializes in the restoration, maintenance, and operation of the Albatross at its facility in California. With a large inventory of spare aircraft for the airframe and engines, the company certifies all former military Albatrosses along with the G-111.

Airpower, Inc. was originally a subsidiary of Transocean Airlines. With a wealth of experience in heavy radial-engined aircraft, this company is in the business of overhauling the Wright powerplant. Located in California, Airpower also owns a former Coast Guard HU-16E numbered 2134. During Hurricane Hugo, the Albatross was blown across the airport, ending up against a hurricane fence, but not before it destroyed several smaller aircraft on its trip across the airfield — damage to the Albatross amounted to scrapes on the rudder and a punctured float. Airpower also purchased two former Mexican Navy Albatrosses it plans to refurbish.

UF-1 131911 trundles out of the Arizona desert in 1988 for its first flight in 25 years. (Photo courtesy Steve Penning)

The sleek lines of N48318 hardly evoke thoughts of the nicknames given the Albatross years earlier by Air Force crews. This former HU-16B, number 51-7187, received its airworthiness certificate in December 1992. (Photo courtesy Ralph Smith)

Reid Dennis, one of the original Silicon Valley venture capitalists, harbors a passion for seaplane aviation matched only by his success in business. He resurrected his Albatross, HU-16C number 137932, from the Arizona desert and transformed it into a virtual winged yacht. Foremost among

Former Navy UF-1 131911 owned by Lynn Hunt and David DeWitt. (Photo courtesy Steve Penning)

Resplendent in its Navy SAR scheme, a renovated Albatross tucks in its landing gear on takeoff in October 1992. (Photo courtesy Steve Penning)

92 • GRUMMAN ALBATROSS: A HISTORY OF THE LEGENDARY SEAPLANE

A pair of former Mexican Navy Albatrosses numbered MP 402 and MP 404, await restoration at La Paz, Baja, California. (Photo courtesy Todd Falconer)

After being fitted with rebuilt engines, UF-1 number 141276 prepares to leave its desert resting spot. The triangle on the bow outlines the insignia of the Icelandic Coast Guard, the aircraft's last duty before storage. (Photo courtesy Steve Penning)

the improvements, far too numerous to list, are 500 pounds of soundproofing insulation, wide-chord nickel-plated props, a rapid fuel dump system, an electrical management center, a six-bilge pump manifold system, turbine APU, modern avionics, and a completely modernized interior featuring observation "bubbles," fully functional galley, complete lavatory with shower, and a compressor for refilling SCUBA tanks. By adding "B" Model wing panels to his shorter UF-1 wings, he essentially created a STOL Albatross.

Much like Reid Dennis, Melvyn Arthur of Arizona long held an avid interest in seaplane aviation to the point where he too resurrected an Albatross from its desert roost. After five years of diligent work, former Navy UF-1 number 137926 was completed. On September 10, 1992, Arthur lifted N888AC from the runway to begin a round-the-world odyssey that retraced historic voyages of exploration.

Albatrosses have also appeared on the civil registers of Brazil, Great Britain, Indonesia, Mexico, Paraguay, and the Philippines. An interesting aspect of the Albatross hulks left derelict in the desert is based on John Wood's contention that there is more nobility in converting doomed airplanes into furniture than beer cans. After successfully transforming other airplane components into furniture, Woods decided to restore an Albatross cockpit to factory-fresh condition and simultaneously end up with a unique desk. The cockpit section was cut from the hulk of TU-16C number 131914, and 800 tedious hours later, emerged as an elegant piece of furniture.

Dennis Buehn's pristine HU-16C is finished in the Navy utility scheme as he remembers it from his Navy days when he flew in 137927. (Photo courtesy Terry Love)

CHAPTER 13: WARBIRDS AND PLEASURE CRAFT • 93

CHAPTER 14

TURBO ALBATROSSES

In its persistence to strive for improvement of the Albatross, Grumman's enthusiasm often met with resistance or outright refusals. As early as 1961, the company offered the U.S. Navy a UF-2 variant powered by General Electric T64 turboprop engines but the Navy would have no part of it. A similar proposal offered to the Air Force during the early 1960s elicited a similar response. Nevertheless, Grumman's optimism would not be dampened as it drew up plans for a turboprop version of the amphibian. Studies were completed in early 1964 which proved the operational and economical feasibility of extending the Albatross' useful life through turbo modification.

Thus, the groundwork had been laid for a "Super Albatross" design tentatively designated the HU-16F, which Grumman planned to build during the summer of 1967 and fly in early 1968. The design called for the installation of two GE T64 turboprop engines rated at 1,425 shp. The proven T64s would allow the Super Albatross shorter takeoffs at heavier weights, improved single-engine performance, greater service ceiling and increased rate of climb, increased speed at all altitudes, and quieter engine operation.

While the HU-16F did not reach fruition, the design evolved into a multi-engine arrangement incorporating four United Aircraft of Canada Model T74 turboprop engines rated at 830 shp each. An adjunct consideration for this variant used alternate 450 gallon droppable fuel tanks in lieu of the standard 300 gallon tanks which would extend search mission duration by nearly two hours. Fitting of the T74s would have required no wing strengthening since the turboprops were lighter and lower powered. Adding the four smoother, quieter engines would actually result in a weight reduction of 2,600 pounds. The propeller diameter of the T74 was 33 inches less than the radial type which broadened the tip-to-water clearance. This would create less spray erosion which was as damaging as gravel impinging on the props. Performance characteristics paralleled those of the twin turboprop with the multi-engine variant boasting a takeoff gross weight of 35,000 pounds.

Further research and development by Grumman in the field of turbo power for the Albatross resulted in plans for another multi-engine variant in the early 1970s. Designated the HU-16BT, this design incorporated four United Aircraft of

Conroy Aircraft Corporation's turbo was powered by a pair of Rolls Royce Dart RDa 6 Mk 510 turboprops designed for the Vickers Viscount. (Photo courtesy Dave Ostrowski)

Grumman artist's rendition of a multi-engine turbo-powered Albatross, circa 1970. (Courtesy Grumman Corporation)

Chalk's first G-11T turbo conversion began life as an Air Force SA-16A which was later transferred to the Coast Guard where it was modified to UF-2G standards. (Photo courtesy Allied Signal)

Canada Model PT6A-34 turboprops rated at 783 shp each. Mounting of the outboard engines did require structural wing reinforcement. One notable feature of the four-engine Albatross was its ability to cruise on two engines which extended the aircraft's range well beyond 2,000 miles. The projected empty weight of the HU-16BT was 21,400 pounds with a maximum takeoff weight of 36,500 pounds.

An offshoot project of turbo-powered Albatrosses took the unique form of the UF-XS developed by the Japanese firm Shin Meiwa. Designed to meet a Japanese Maritime Self Defense Force (JMSDF) requirement, the highly modified variant also served as a scaled down proof-of-concept aircraft for Shin Meiwa's PS-1 flying boat. The project started out as a UF-2 assigned USN Bureau Number 149822, which was reserialed 9911 by the JMSDF. The UF-XS made its first flight on December 20, 1962, and underwent operational evaluation with the JMSDF Albatross unit during 1963 and 1964.

The UF-XS was structurally altered to the point where it barely resembled an Albatross. It was given a lengthened hull, deeper afterbody, T-tail and swept fin empennage, slatted wings, and a new undercarriage. The complex mix of powerplants included the original Wright radials with a pair of Pratt & Whitney R-1340-AN-10 radials added outboard. Two 1,000 hp GE T58-GE-6 turbines, housed atop the fuselage, supplied compressed air for the wing boundary-layer control system. The aircraft had a maximum takeoff weight of 35,500 pounds and a maximum speed of 207 mph. It was later put on permanent display at Shimofusa.

Another turbo Albatross conversion was undertaken by the Conroy Aircraft Corporation at Santa Barbara, California, in 1969. Originally a SA-16A with serial number 51-004, the Albatross was later registered as N459U which became N16CA under Conroy ownership in September 1969. This turbo variant mounted a pair of Rolls Royce Dart engines and made its first flight in February of 1970. The aircraft is presently kept at Conroy's facility at Salisbury, Maryland. In February 1992, Conroy acquired a second Albatross, registered N16ZE, which saw service with the Coast Guard as HU-16E number 2124.

Grumman conversion of the turbo-powered Albatross was again on the drawing boards in 1983. In its quest to re-engine the amphibian, Grumman set its sights on Garrett's 1,645 shp TPE331-15UAR turbo engine. While the Wright radials consistently performed effectively on the Albatross, there was no denying the advantages of a modern engine. Combined with a low-noise Dowty Rotol four-bladed propeller, installation of the TPE331 would reduce powerplant weight by more than 2,000 pounds and allow greater stability and single-engine performance.

Other improvements with the new engines realized a maximum takeoff weight of 37,500 pounds which translates to doubling the cargo capacity to eight tons. The Pratt & Whitney PT7 and the GE CT7 were also considered as replacements for the radials.

In mid 1993, Chalk's International Air lines, through Flying Boat Incorporated, entered the initial stages of a plan to convert their fourteen Albatrosses to turbo power with the Allied Signal/Garrett TPE331-14GR engine rated at 1,650 shp. One G-111, registered N120FB, has been modified to a G-111T with additional units planned for availability in 1995. The conversion would take the Albatross into the 21st century and, in conjunction with foreseeable Cuban routes re-opened, allow Chalk's to put their G-111Ts into full operation.

CHAPTER 14: TURBO ALBATROSSES • 95

CHAPTER 15

MUSEUM AND DISPLAYS

Some Albatrosses were spared the fate that befell many of the amphibians during the 1970s. A final resting place in the Arizona desert marked the end for many Albatrosses that made their final flight to the Military Aircraft Storage and Disposition Center (MASDC), now called the Aerospace Maintenance and Regeneration Center (AMARC). More than twenty Albatrosses had their careers extended by serving as monuments in worldwide tribute to the amphibian's impeccable service. Eighteen examples alone are proudly displayed at museums and military installations across the country, while others are museum residents in Europe, South America, and China.

Included among these displays are Albatrosses that distinguished themselves in one way or another. The HU-16B displayed at the U.S. Air Force Museum at Wright-Patterson AFB, Ohio, for example, was flown there in mid July 1973 by Col. Charles Manning who two weeks earlier piloted the aircraft on a record altitude flight. Number 51-5282 was the last Air Force Albatross, having served with the 301st ARRS at Homestead AFB, Florida.

Another 301st HU-16B was also earmarked for preservation in 1973. Still in the Air Rescue scheme, number 51-7144 was delivered to the Museum of Aviation at Warner Robins AFB, Georgia, where it took its place among more than 85 historic aircraft. In 1965 the Air Logistics Center at Warner Robins assumed worldwide logistics management responsibility for the HU-16.

HU-16B number 51-7200 arrived at Chanute AFB, Illinois, in December 1974 where it became an impressive roadside display. This Albatross underwent extensive modifications during the 1950s which transformed it from an SA-16A into the first highly improved "B" Model.

When the aircraft carrier USS INTREPID started her second career as a museum in New York City, the Coast Guard seized the opportunity to place an Albatross aboard. The aircraft selected for the display, HU-16E number 7216, was in disrepair after ten years of idle exposure to the elements. With the combined efforts of the Coast Guard Auxiliary from the 3rd Northern District and regular USCG personnel from Brooklyn, a restoration program was underway in April 1983.

The HU-16B, serial number 51-7163, on display at the Castle AFB Museum is in pristine condition. (Photo courtesy Castle Air Museum)

Left: This Albatross, on display at the Air Rescue Museum at Kirtland AFB, New Mexico, was originally a UF-2G which was donated by the Coast Guard. It is marked as HU-16B 51-071 to memorialize the Albatross which was lost in combat off the coast of North Vietnam. (Photo courtesy David Wendt) Right: HU-16E number 2129 at the USS ALABAMA Memorial Park. (Photo courtesy Terry Love)

96 • GRUMMAN ALBATROSS: A HISTORY OF THE LEGENDARY SEAPLANE

Left: The Albatross at the Pima Air and space Museum at Tucson, Arizona, is the only known SA-16A on display in the U.S. Serial number 51-022 wears its original rescue markings. (Photo courtesy Chuck Pomazal) Right: HU-16B 51-5303 at Lackland AFB, Texas, is finished in the later overall gray scheme. (Photo courtesy Terry Love)

Above and below: The HU-16B displayed at the Robins AFB, Georgia, Museum is complete down to the fuel tanks, JATO mounts and rescue scheme. Serial number 51-7144 has a long history including rescue service in Vietnam. 51-7144 carries the MAC emblem on the tail fin which reveals blue paint from the camouflage scheme worn in Vietnam. Unusual is the sea rescue platform in place. (Photos courtesy of Darwin Edwards)

After nearly eighteen months, the Albatross was completed and hoisted aboard.

The National Museum of Naval Aviation at NAS Pensacola, Florida, displays a USCG HU-16E number 7236. This Albatross was received in flying condition from Coast Guard Air Station Traverse City, Michigan, in October 1977.

During the late 1970s, the New England Air Museum at Bradley International Airport, Windsor Locks, Connecticut, received HU-16B number 51-025. After being destroyed by a tornado, this aircraft was replaced by HU-16E number 7228 received from the Coast Guard in 1983. It resides with five other Grumman aircraft as part of a collection of nearly 130 aircraft. Its last active base was Miami where helicopters replaced the venerable Albatross.

The Pate Museum of Transportation at Ft. Worth, Texas, acquired HU-16B number 51-7176 from the U.S. Air Force Museum loan program in 1971. It was flown to the Pate facility from Warwick, Rhode Island, where it served with the 143rd Special Operations Group.

Of all the Albatrosses on display, the only original SA-16A is found at the Pima Air and Space Museum at Tucson, Arizona. There, number 51-022 takes its place among twelve other Grumman aircraft, including a "Widgeon", in a collection of nearly 300 historic aircraft. One other SA-16A was briefly exhibited at the Michigan Military and Air Museum at Saginaw, Michigan, before the venture went bankrupt.

Typical of the many Albatrosses that soldiered on through multiple careers, HU-16B number 51-7251 began life as an Air Force rescue craft in December 1953. It was transferred to the Coast Guard in February 1959 and served until July

HU-16B 51-025 at the New England Air Museum. The anti-collision light incorporated into the base of the tail fin was added during "B" Model modification. (Photo courtesy Terry Love)

HU-16B 51-7176 at the Pate Museum carries 300 gallon fuel tanks which was the standard size used by the Air Force. (Photo courtesy the Pate Museum)

CHAPTER 15: MUSEUM AND DISPLAYS • 97

HU-16B 51-7251 was delivered to the Air Force in December 1953 and was transferred to the Coast Guard in February 1959. It was retired in July 1982 and placed on display in USAF livery at the Linear Air Park, Dyess AFB, Texas. (Photo courtesy SGT Carl L. Cook, USAF)

Resplendent in its Coast Guard colors, HU-16E number 7236 is displayed at the National Museum of Naval Aviation at Pensacola as part of an extensive Coast Guard collection. (Photo courtesy National Museum of Naval Aviation)

1982. It was then turned over to the Linear Air Park at Dyess AFB, Texas, where it again donned an Air Force uniform as a monument sponsored by the 40th Airlift Squadron.

Originally an SA-16A, HU-16B number 51-7209 was delivered to the Air Force on July 16, 1953. After a mere seven years of service divided between two units, it was placed in storage but transferred to the Coast Guard one year later. In October 1978 it was flown to Luke AFB, Arizona, for display and on March 28, 1988, it was carried by a C-5 to McClellan AFB, California, for restoration and display in the Aviation Museum.

Rare as they were, an Albatross utilized by the USAF Strategic Air Command (SAC) was acquired by the SAC Museum at Bellevue, Nebraska, near Offutt AFB. Adorned with the distinctive SAC band and emblem, HU-16B number 51-006 serves as a prominent display.

HU-16B number 51-7163 ended up at the Castle AFB Museum in California after an extensive career which included its use as a commercial carrier with Pan American World Airways under U.S. government contract.

In early June 1978, the 1550th Aircrew Training and Test Wing (ATTW) at Kirtland AFB, New Mexico, dedicated an HU-16B which was incorporated into the Rescue Museum. The Coast Guard donated the UF-2G, number 1280, which was flown from the Mobile Air Station. As a memorial to the Air Rescue HU-16B which was blasted from the waters of North Vietnam with the loss of two airmen, the amphibian was repainted in USAF livery and given the serial number of the lost Albatross, 51-071.

Originally an Air Force amphibian, HU-16E number 51-7254 last served the Coast Guard flying from Cape Cod Air Station, Massachusetts. On May 25, 1979, it was flown into storage where it spent the next decade sitting in the Arizona sun. Its fate with the smelter was spared when it was flown to the Travis AFB Museum where it was painted in Air force markings and placed on display.

HU-16Bs are also displayed at the Florence Air and Missile Museum, South Carolina (51-7212); Lackland AFB, Texas (51-5303); and at the Maryland ANG base at Glenn L. Martin Airport (51-7193). Coast Guard HU-16E displays include: number 2129 at the USS ALABAMA Memorial Park, number 7250 at Coast Guard Air Station Cape Cod, Massachusetts, and number 7247 at Air Station Elizabeth City, North Carolina.

This Albatross was assigned to SAC early in its career. It began service on September 26, 1951, with the Air Force where it served the 2nd, 3rd and 33rd Air Rescue Squadrons. It was assigned to the 143rd Airlift Squadron in November 1956 and converted to a SA-16B in June 1957. Number 51-006 commenced duty with the Strategic Air Command on November 10, 1958, with assignment to SAC's 22nd Helicopter Squadron. Thereafter, it also served the 4082nd and 95th Strategic Wings, both at Goose AB, Newfoundland, before it was finally declared surplus in May 1970 and sent to the SAC Museum near Offutt AFB, Nebraska. (Photo courtesy Mark Trupp)

Number 51-7209 was delivered to the Air Force on July 16, 1953, as a SA-16A and assigned to the 1707th Training Squadron. In May 1959 it was transferred to Brookley AFB, Alabama, and one year later, was placed in storage. It went to the Coast Guard in April 1961 where it operated from three air stations until retired to Luke AFB for display. In March 1988 it was carried by C-5 to McClellan AFB, California, where it rests in beautifully restored condition. (Photo courtesy Jim Dunn)

Following extensive restoration, HU-16 7216 joined the impressive aircraft display aboard the USS INTREPID Museum docked in New York City. (Photo courtesy Frank DeSisto)

100 • GRUMMAN ALBATROSS: A HISTORY OF THE LEGENDARY SEAPLANE

In a park-like setting, Chanute AFB, Illinois, displaced the first Albatross converted to a HU-16B. Number 51-7200 wears Air Rescue markings. (Photo courtesy Lennart Lundh)

The New England Air Museum owns this beautifully maintained HU-16E, number 7228, formerly assigned to CGAS Miami. (Photo courtesy Lennart Lundh)

Left: The last operational Air Force Albatross, HU-16B number 51-5282, lands in Biscayne Bay after its record-breaking altitude flight in July 1973. Afterwards, it was flown to the Air Force Museum at Wright-Patterson AFB, Ohio. (Photo courtesy U.S. Air Force Museum via Dave Menard) Right: HU-16B 51-5303 at the History and Traditions Museum at Lackland AFB, Texas. (Photo courtesy History and Traditions Museum)

CHAPTER 15: MUSEUM AND DISPLAYS • 101

LINE SCHEMES

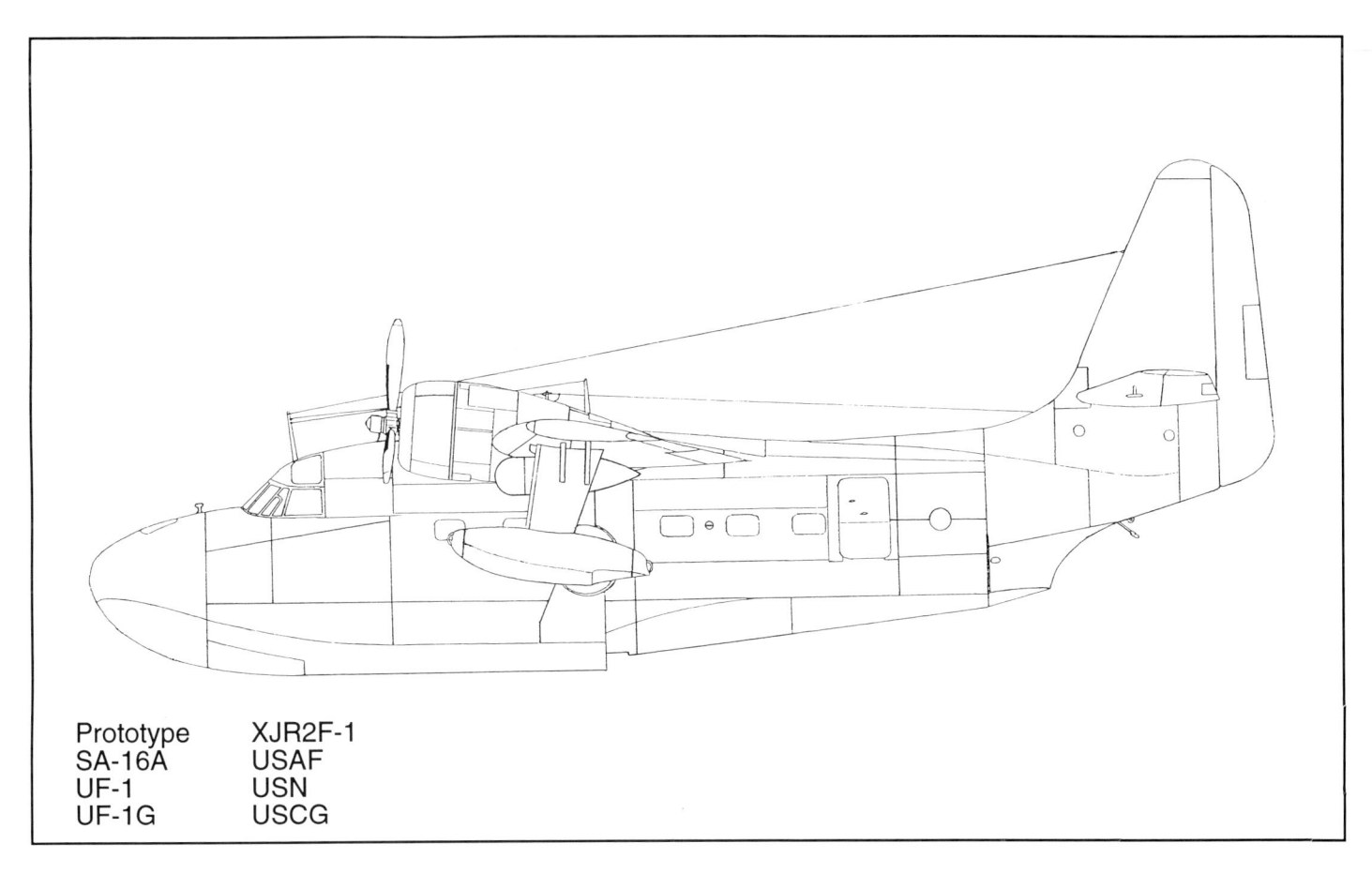

Prototype   XJR2F-1
SA-16A   USAF
UF-1   USN
UF-1G   USCG

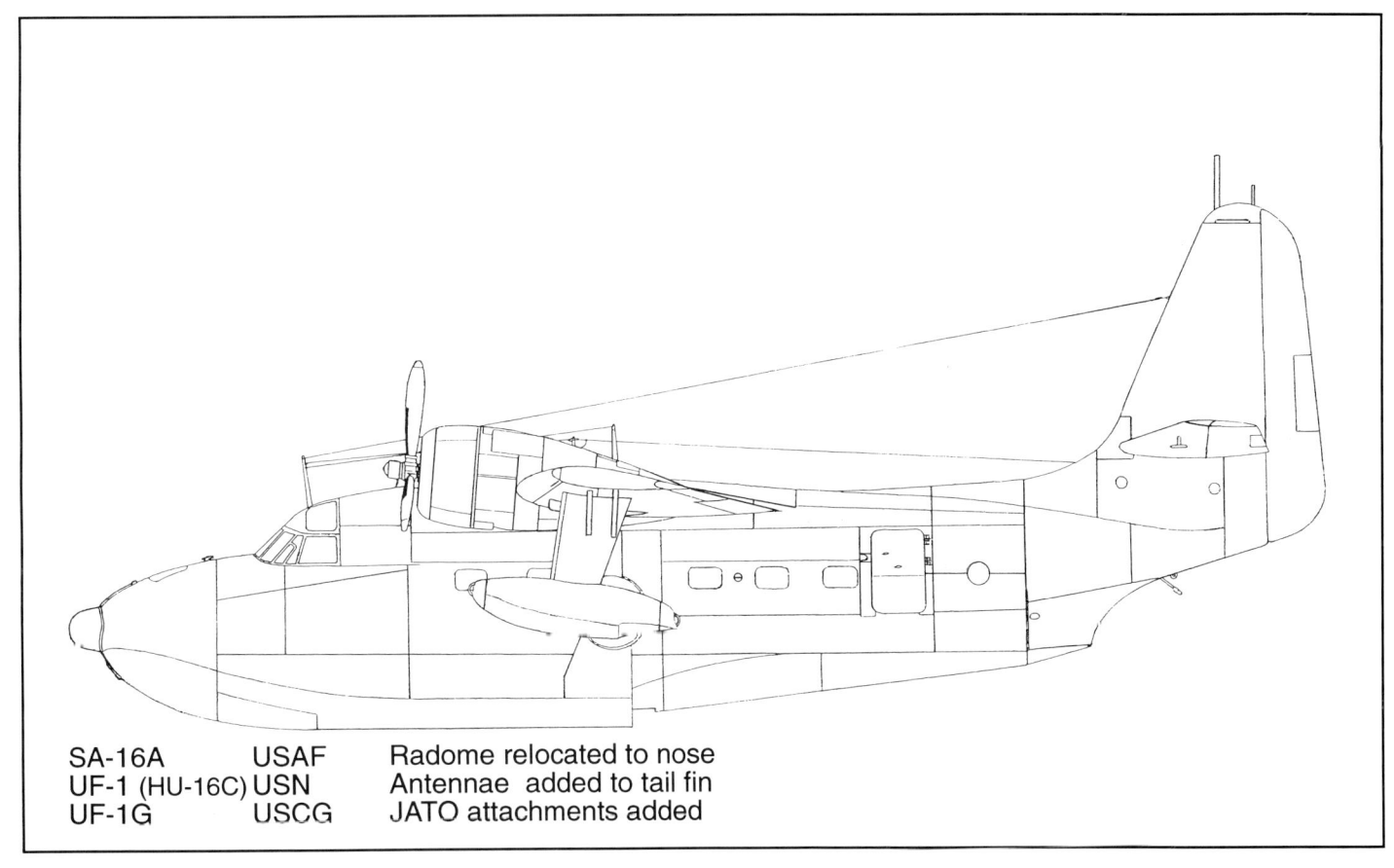

SA-16A   USAF   Radome relocated to nose
UF-1 (HU-16C)   USN   Antennae added to tail fin
UF-1G   USCG   JATO attachments added

LINE SCHEMES • 103

SA-16A Triphibian

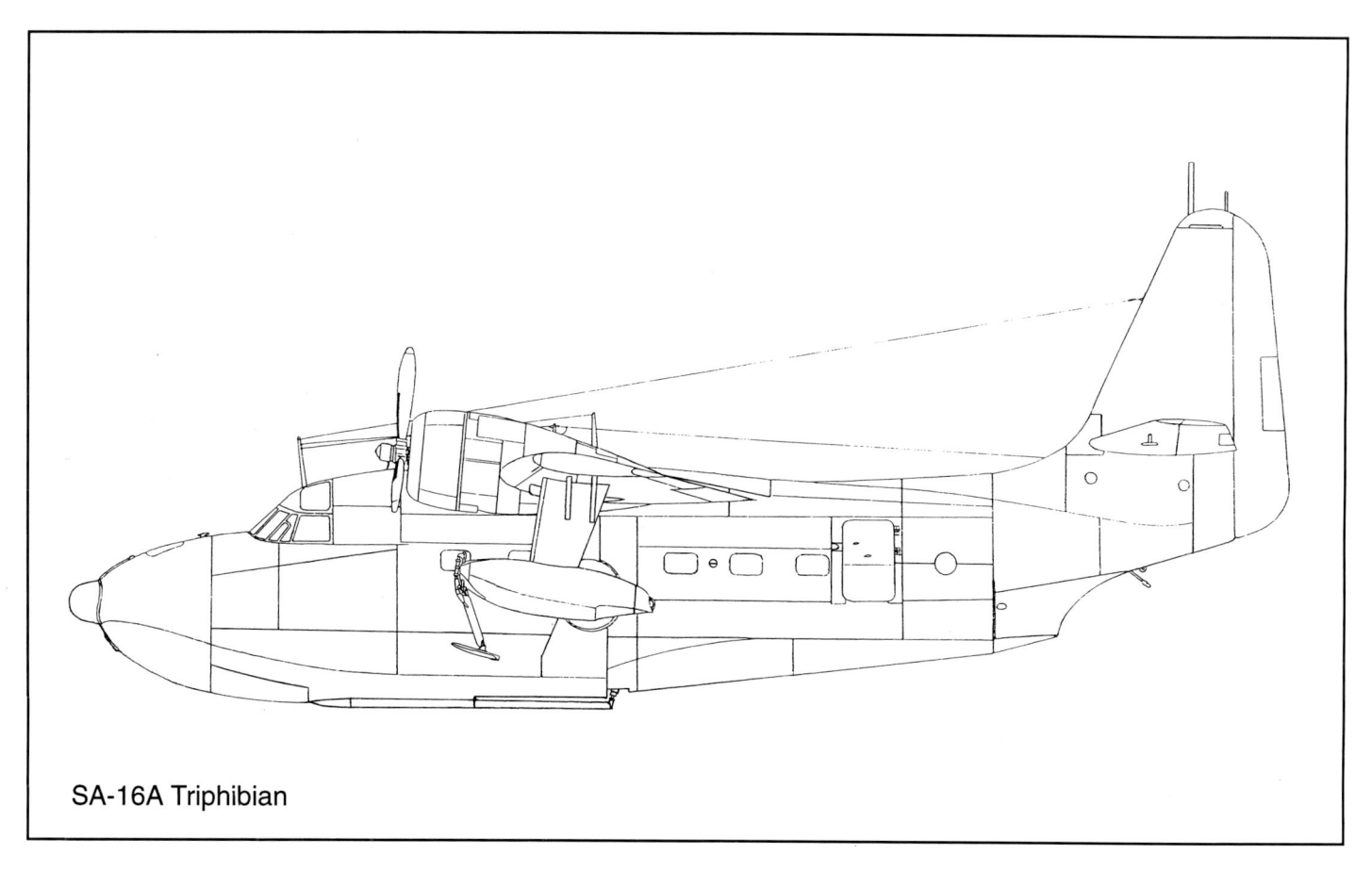

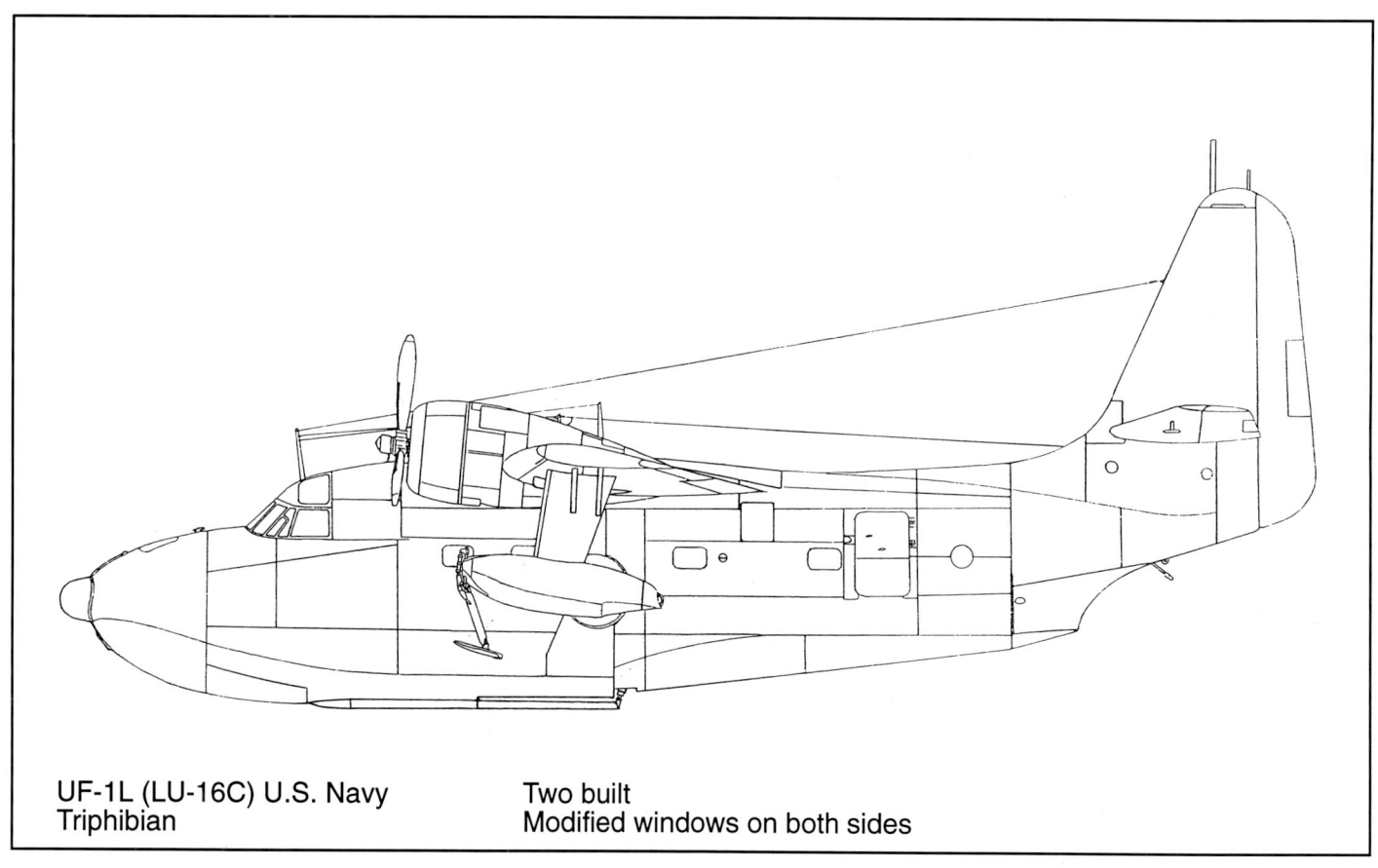

UF-1L (LU-16C) U.S. Navy Triphibian   Two built
Modified windows on both sides

104 • GRUMMAN ALBATROSS: A HISTORY OF THE LEGENDARY SEAPLANE

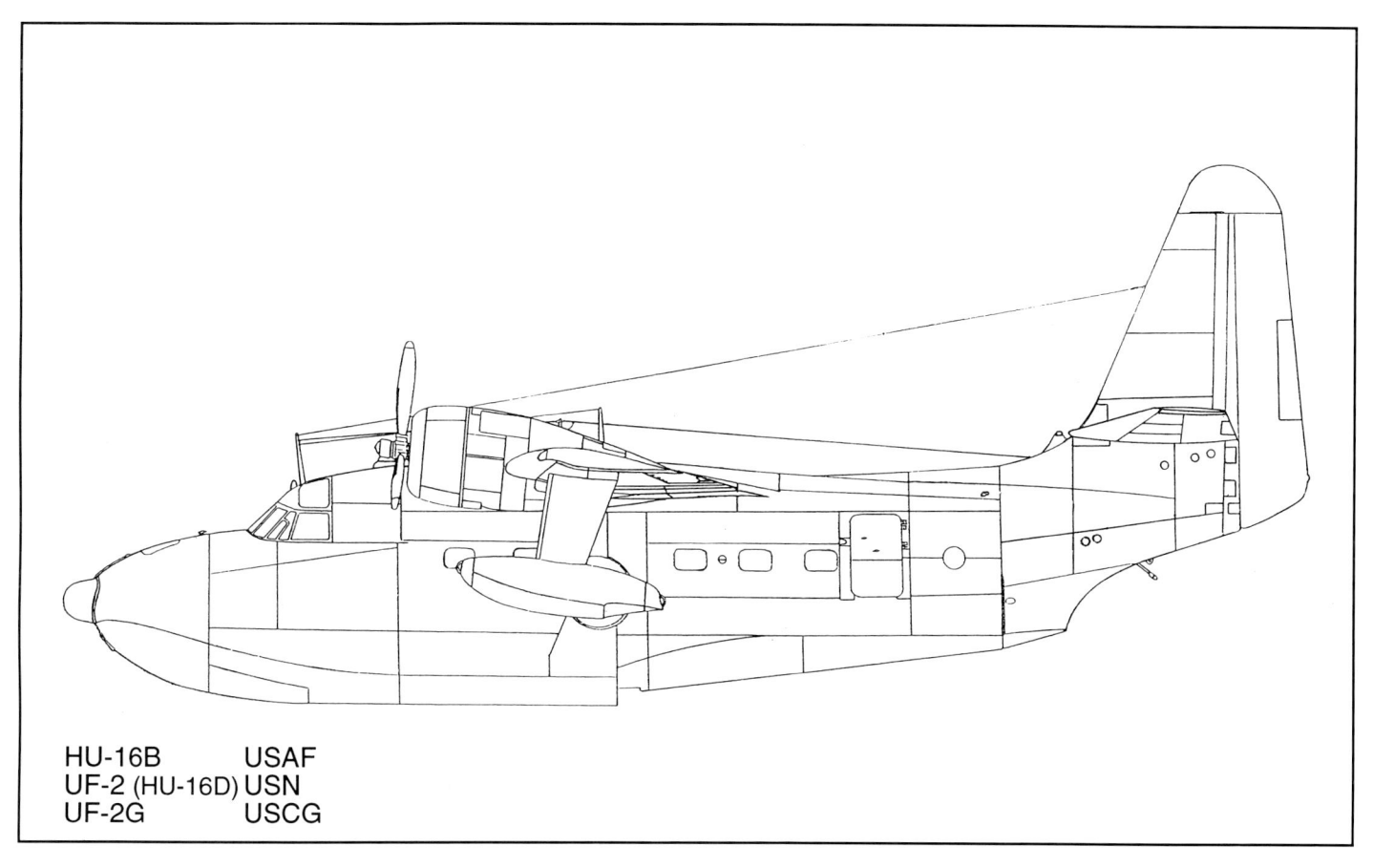

HU-16B USAF
UF-2 (HU-16D) USN
UF-2G USCG

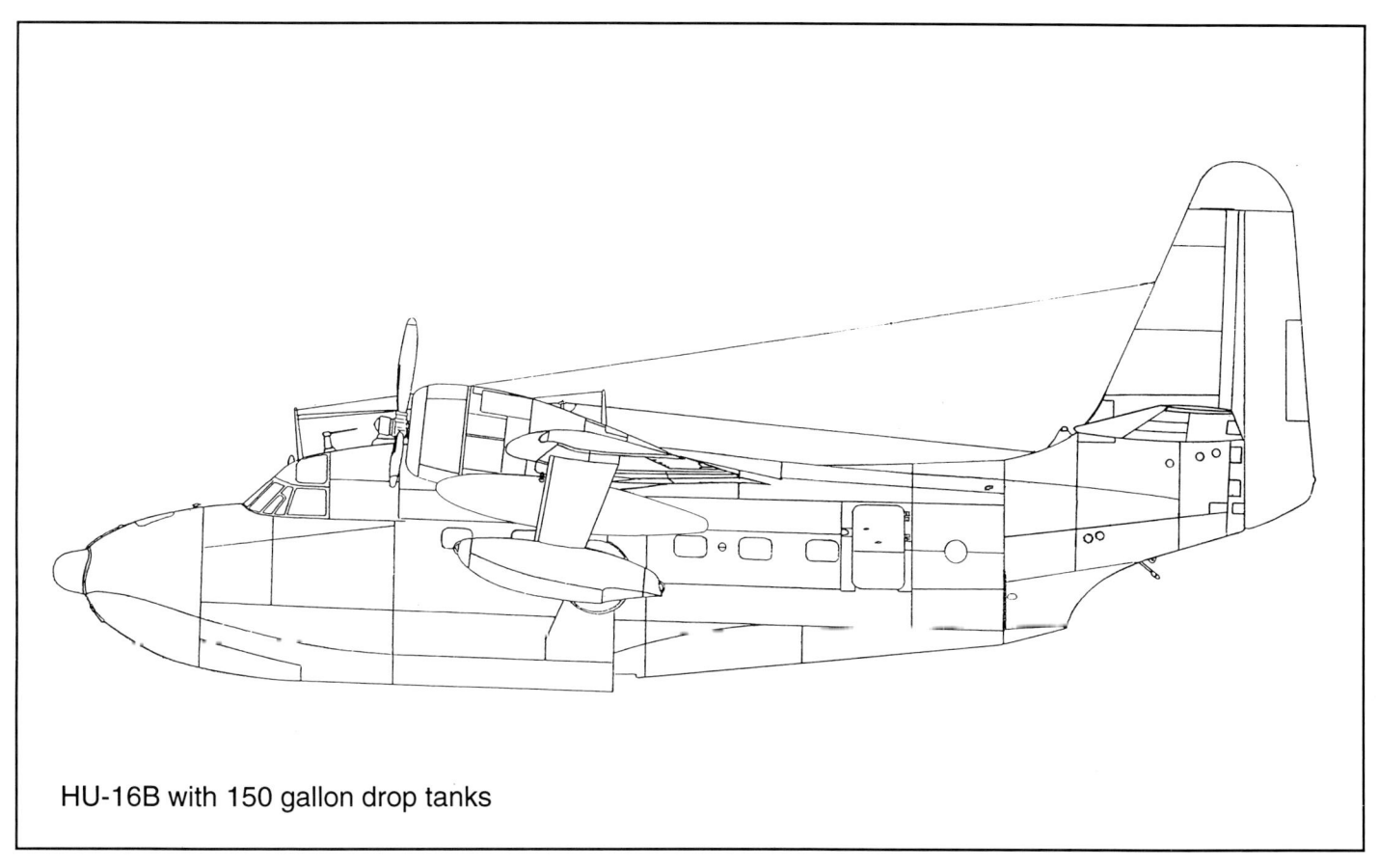

HU-16B with 150 gallon drop tanks

LINE SCHEMES • 105

HU-16B with 300 gallon drop tanks

SHU-16B ASW Version
Modified nose and hull
MAD boom

Upper wing antennae added
Underwing stores stations

106 • GRUMMAN ALBATROSS: A HISTORY OF THE LEGENDARY SEAPLANE

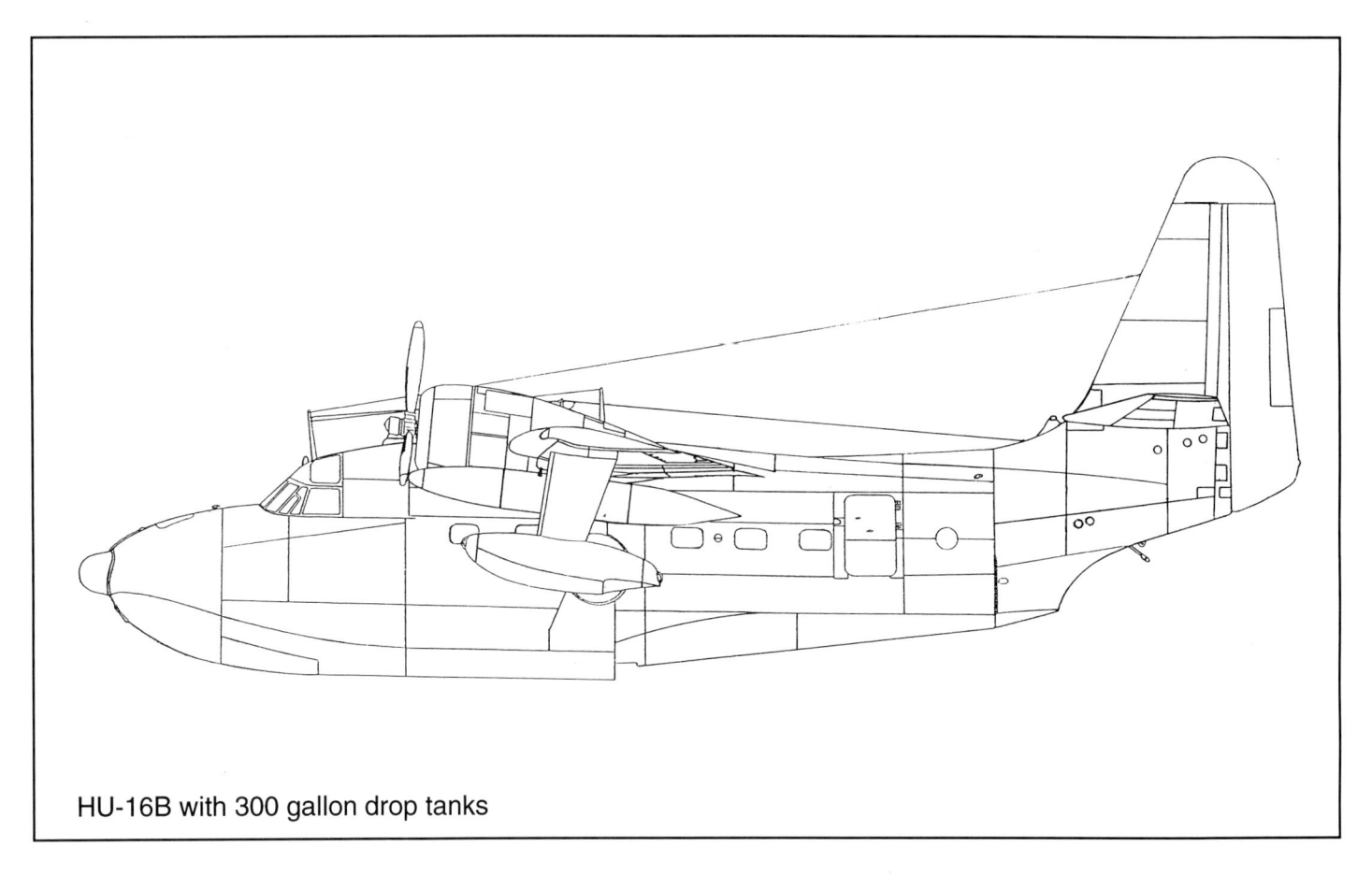

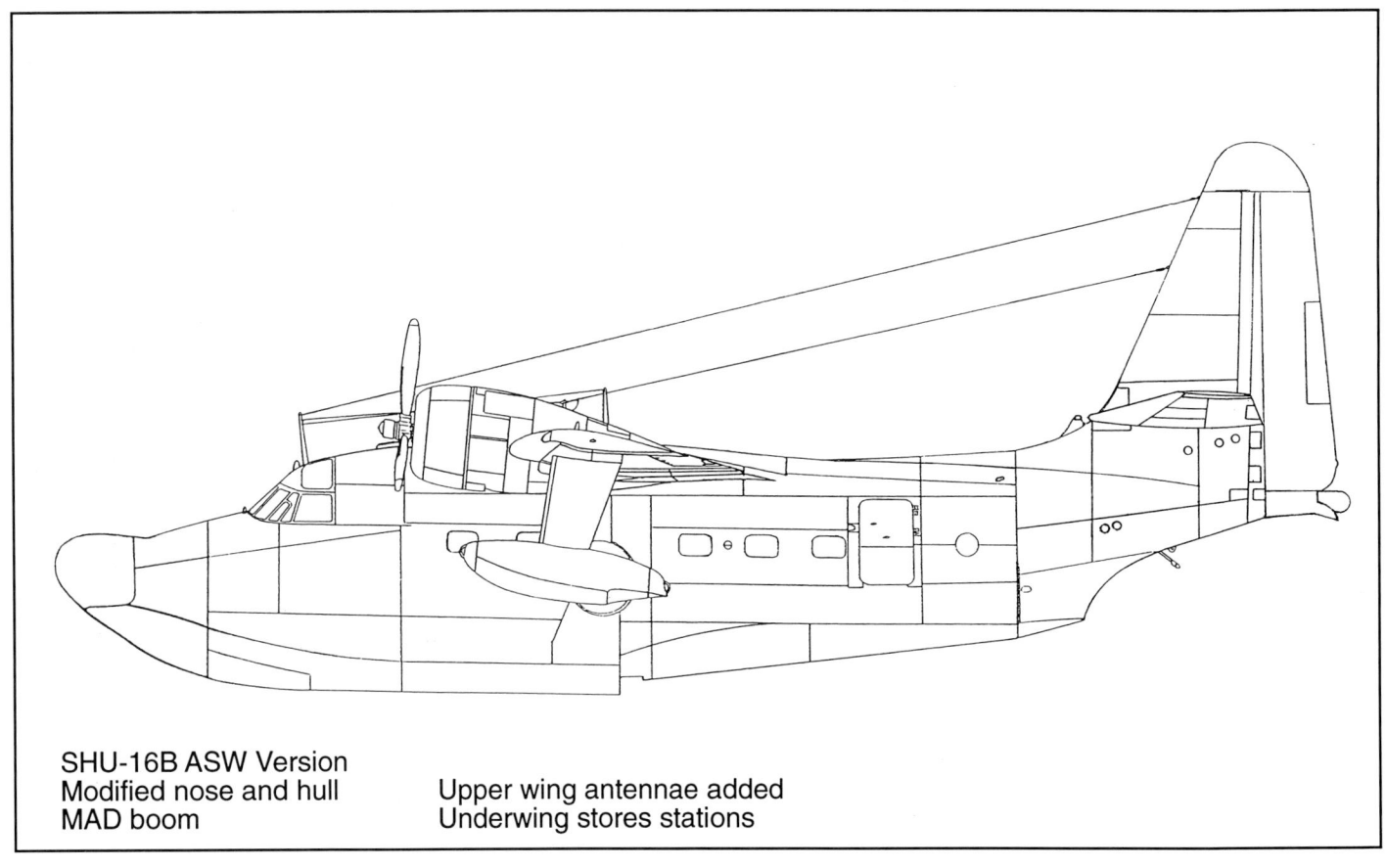

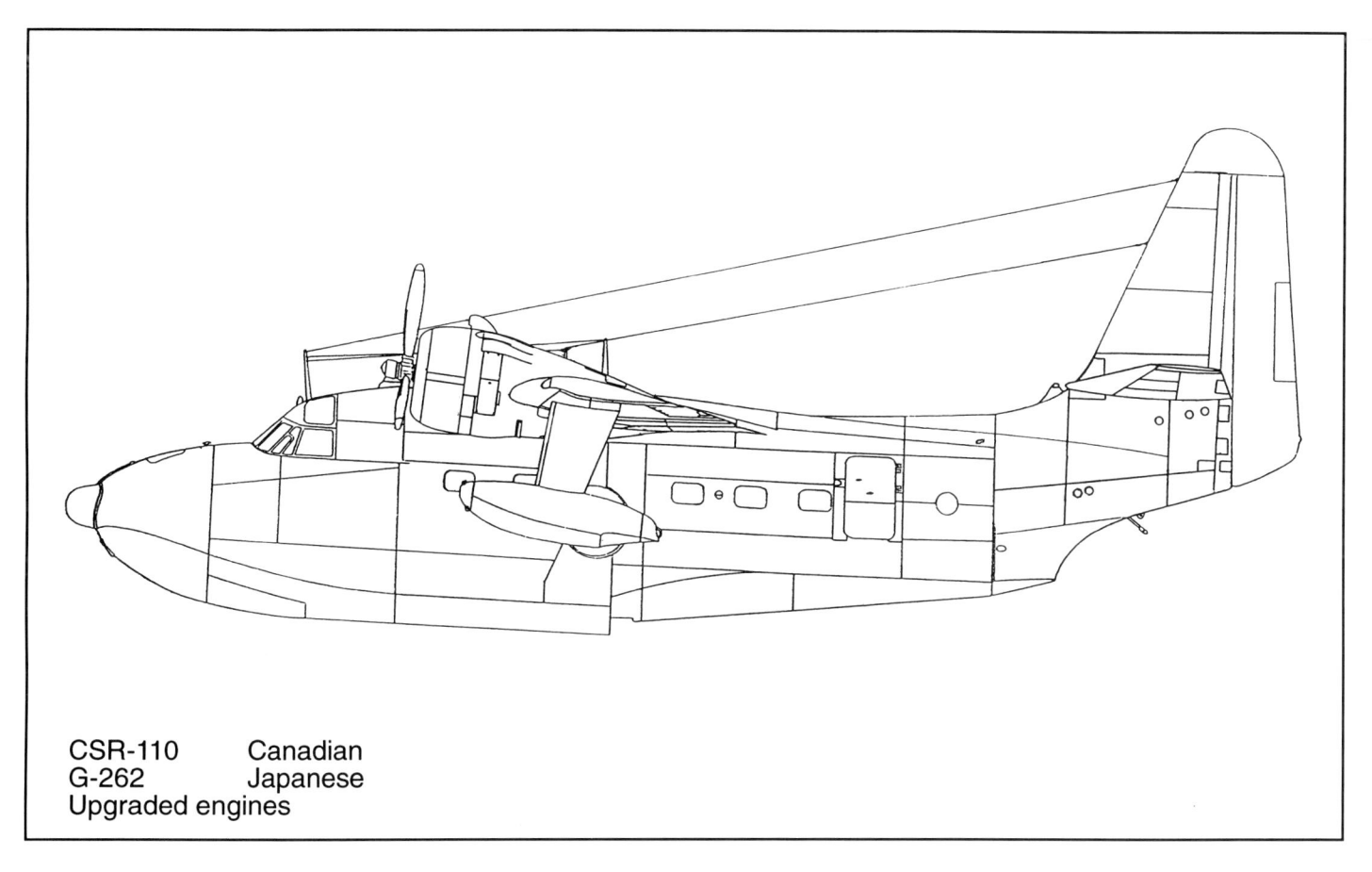

CSR-110   Canadian
G-262   Japanese
Upgraded engines

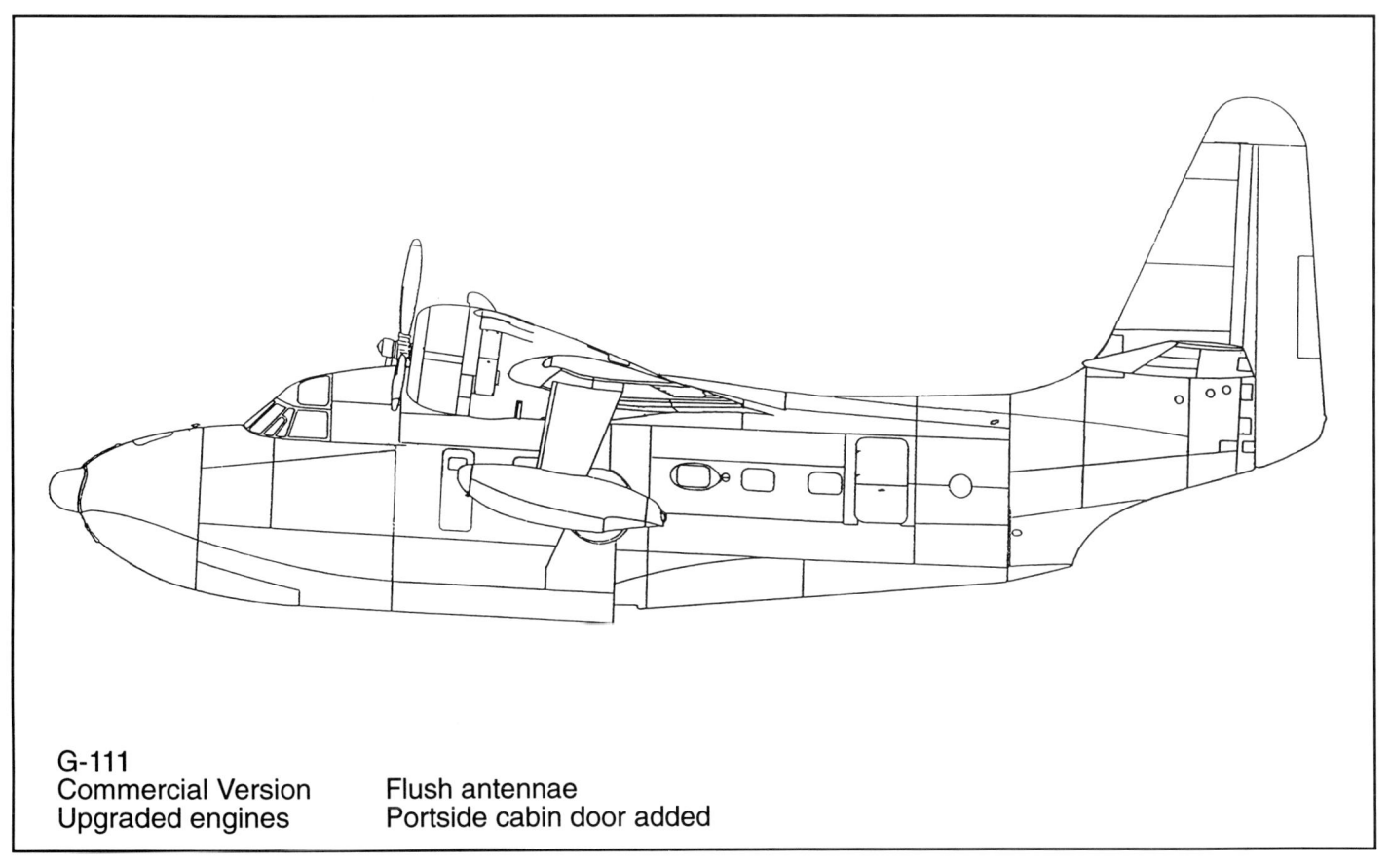

G-111
Commercial Version   Flush antennae
Upgraded engines   Portside cabin door added

LINE SCHEMES • 107

Also from the publisher

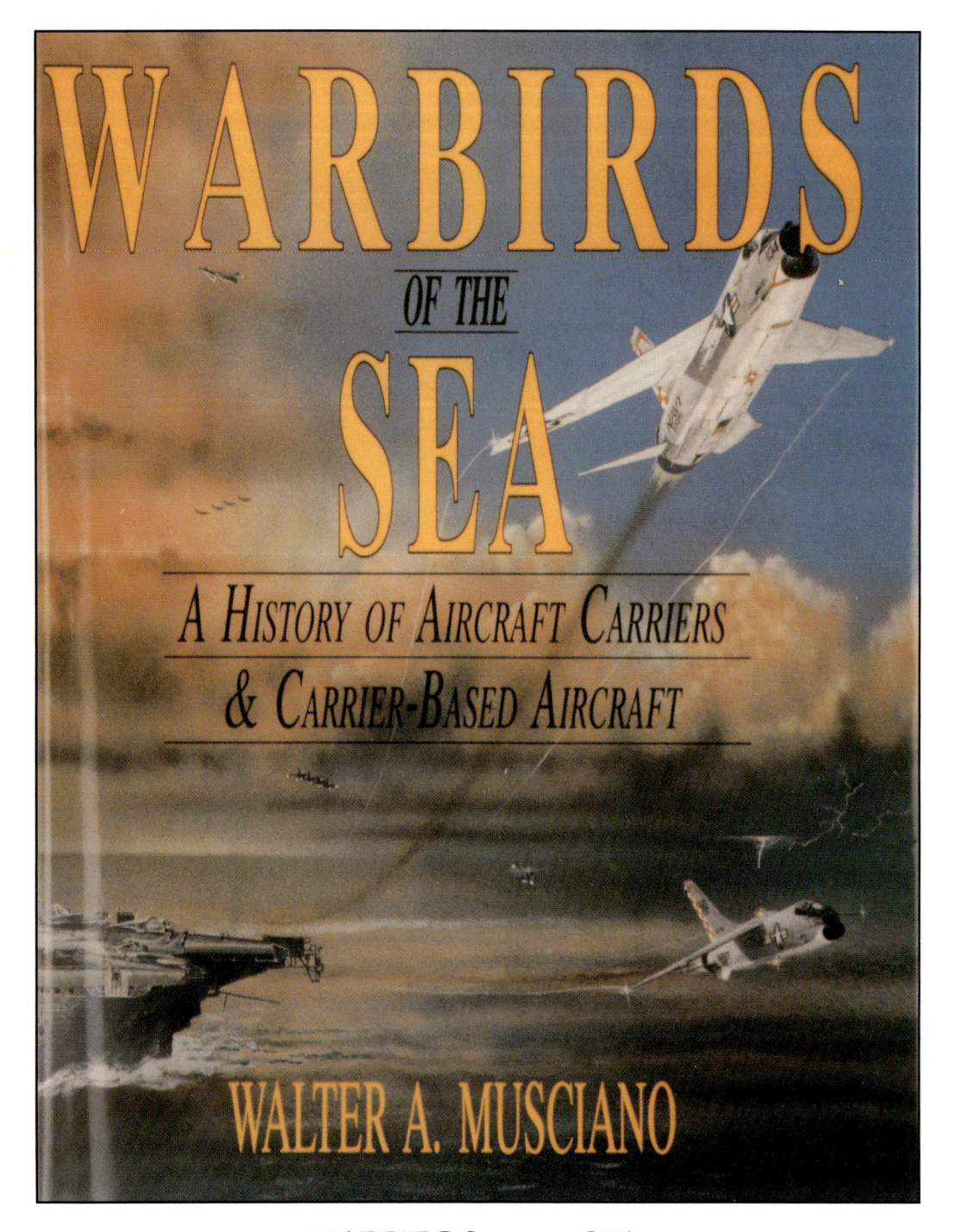

WARBIRDS OF THE SEA
A History of Aircraft Carriers & Carrier-Based Aircraft

Walter A. Musciano

Covers the history and combat career of aircraft carriers and shipboard aircraft from their conception into the future.
Size: 8 1/2" x 11" over 800 b/w photos, drawings, maps
592 pages, hard cover
ISBN: 0-88740-583-5 $49.95

NAVAIR 01-245FDD-1
NATOPS FLIGHT MANUAL

Navy Model
F-4J
AIRCRAFT

A SCHIFFER MILITARY HISTORY BOOK

NATOPS Flight Manual
F-4J

This facsimile reprint is the actual NATOPS flight manual for the F-4J Phantom II U.S. Navy fighter aircraft. Operational aspects covered are: flight characterstics, emergency procedures, all-weather operations, communications procedures, weapons systems, flight crew coordination, NATOPS evaluation, and performance data.
Size: 8 1/2" x 11" diagrams, charts, illustrations, appendices, index
544 pages, soft cover $39.95

UNITED STATES COMBAT AIRCREW SURVIVAL EQUIPMENT
WORLD WAR II TO THE PRESENT
A REFERENCE GUIDE FOR COLLECTORS
Michael S. Breuninger

This new book is a detailed study of United States Air Force, Army, Army Air Force, Navy, and Marine Corps aircrew survival equipment. Among the items covered are: survival vests, leggings, and chaps, life preservers, survival (ejection) seat and back pad kits, personal survival kits and first aid kits, and various survival components such as radios, strobes and signal lights, smoke and illumination flares, flare launchers, signal mirrors, rations, miscellaneous food procurement items, desalting kits, fire starter and match cases, survival rifles, survival tools and knives, compasses, survival manuals, personnel lowering devices, one-man life rafts, blood chits, escape maps, and phrase books. Tag and label information is provided for each item
Size: 8 1/2" x 11" over 170 b/w photographs, drawings
208 pages, soft cover
ISBN: 0-88740-791-9 $29.95

DOUGLAS A-1 SKYRAIDER
A Photo Chronicle
Frederick A. Johnsen

The famed Skyraider in Korea and Vietnam, emphasizing its great ground assault capabilities.
Size: 8 1/2" x 11" over 100 b/w, and color photographs
112 pages   soft cover
ISBN: 0-88740-512-6 $19.95

TAIL CODE
USAF: The Complete History of USAF Tactical Aircraft Tail Code Markings
Patrick Martin

Full color history covers PACAF, TAC, AFRES, ANG, AAC, USAFE codes and markings.
Size: 8 1/2" x 11"   240 pages   hard cover, over 300 color photos
ISBN: 0-88740-513-4 $45.00

FORCE DRAWDOWN
A USAF Photo History 1988-1995
René Francillon/Jim Dunn/Carl E. Porter

Illustrated with over 410 color photos, this new book provides a rich pictorial record of aircraft (including old and new markings) and units which no longer exist, and offers a visual chronicle of organizational changes between 1988 and 1995.
Size: 8 1/2" x 11" over 410 color photographs
144 pages, hard cover
ISBN: 0-88740-777-3 $29.95

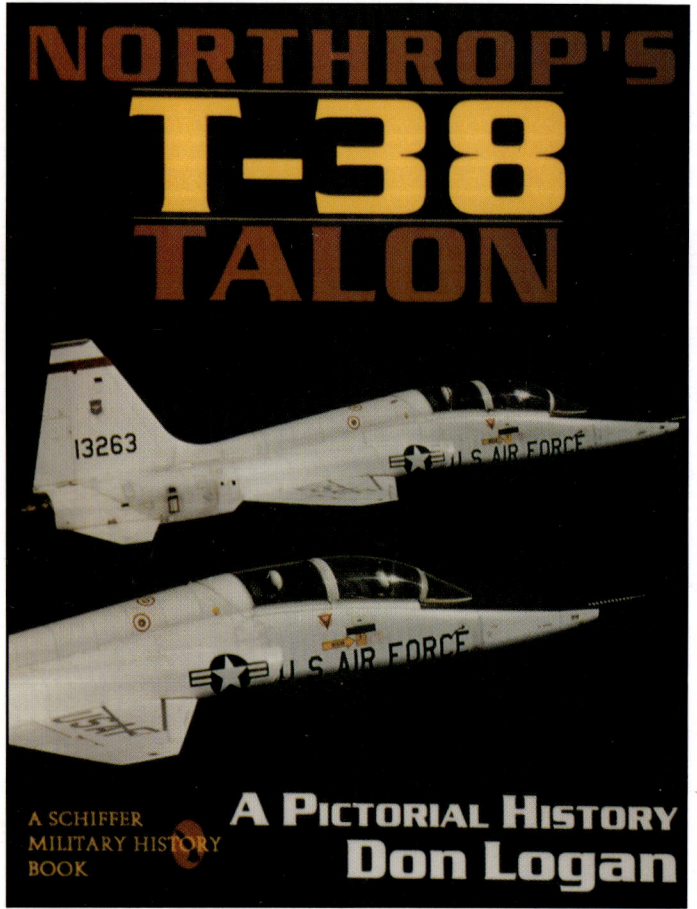

NORTHROP'S T-38 TALON
A PICTORIAL HISTORY
Don Logan

This is the story of the most successful pilot training jet ever produced: the Northrop T-38 Talon. The history of the aircraft is broken down by the roles it has played in over thirty years of service including development and testing, pilot training, flight test support, NASA program support, air combat aggressor, aerial target, Thunderbird-USAF air demonstration team aircraft, companion trainer, and civilian test support. All units flying the T-38, their markings and paint schemes are covered in over 300 color photographs – including a chart of the colors used listing Federal Standard (FS) color numbers. Don Logan is also the author of *Rockwell B-1B: SAC's Last Bomber*, and *The 388th Tactical Fighter Wing: At Korat Royal Thai Air Force Base 1972* (both titles are available from Schiffer Publishing Ltd.).
Size: 8 1/2" x 11" over 300 color photographs
152 pages, soft cover
ISBN: 0-88740-800-1 $24.95

THE 388TH TACTICAL FIGHTER WING
AT KORAT ROYAL THAI AIR FORCE BASE 1972
Don Logan

This new book covers the 388th TFW; a Composite Wing based at Korat RTAFB, Thailand, consisting of fighters, Wild Weasel aircraft, airborne jamming aircraft and AWACS aircraft. The author flew 133 combat missions in Southeast Asia in 1972, and was assigned to the 469th TFS, one of the two F-4E squadrons of the 388th TFW. The book discusses in detail the Wing, the Squadrons and the aircraft they flew: the F-4, F-105G Wild Weasel, A-7D, EB-66, EC-121, and C-130. Also covered are the mission types, as well as operations of the Wing during the Linebacker Campaign over North Vietnam. Narratives of all the 388th MiG kills and aircraft losses during 1972 are included. The book contains over 170 color and black and white photographs taken by the author, as well as theatre maps. A selection of official and unoffical flight suit patches is also included. Don Logan is also the author of *Rockwell B-1B: SAC's Last Bomber*, and *Northrop's T-38 Talon: A Pictorial History* (both titles are available from Schiffer Publishing Ltd.).
Size: 8 1/2" x 11" over 170 color and b/w photographs, maps
128 pages, hard cover
ISBN: 0-88740-798-6 $29.95